KU-018-574

# TWENTY FIRST CENTURY
# SCIENCE

## Project Directors

Angela Hall          Emma Palmer

Robin Millar         Mary Whitehouse

## Editors

Angela Hall

Mary Whitehouse

Emma Palmer

## Authors

| | | | |
|---|---|---|---|
| Colin Bell | Byron Dawson | Caroline Shearer | Carol Usher |
| Jenifer Burden | Andrew Hunt | Carol Tear | Vicky Wong |
| Peter Campbell | Carol Levick | Charles Tracy | |

THE UNIVERSITY *of York*

THE SALTERS' INSTITUTE

Nuffield Foundation

OCR
RECOGNISING ACHIEVEMENT

OXFORD
UNIVERSITY PRESS

Official Publisher Partnership

# Contents

# Features of all living things

① **a** The diagram below shows the processes carried out by all living things.

Make notes in each of the boxes to describe and explain the reason for each process.

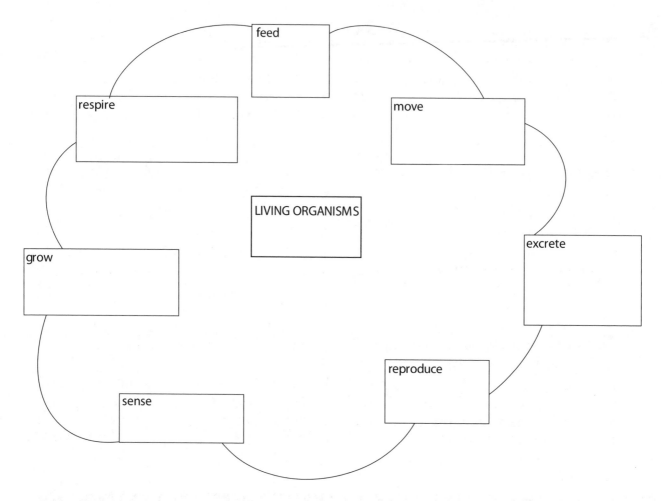

feed

respire

move

LIVING ORGANISMS

grow

excrete

sense

reproduce

**b** Write down the names of two chemical processes that occur in all green plants.

1 .................................................................................................................................

2 .................................................................................................................................

# Enzymes

## (1) Enzymes

Enzymes are a very important group of chemicals found in living things.

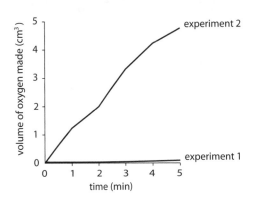

**a** Which experiment used catalase?

........................................................................................

**b** Explain your answer to part a.

........................................................................................

**c** Complete the boxes to explain how the enzyme catalase works. Then draw the appropriate diagram in the box.

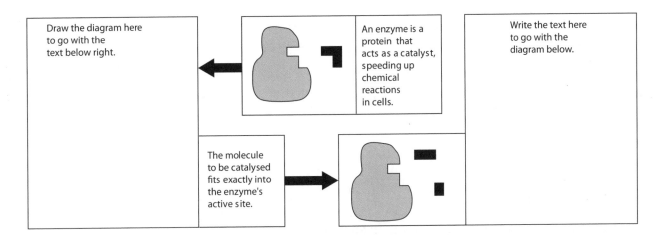

**d** There are tens of thousands of enzymes in the human body. Each one speeds up a different chemical reaction.

Explain why an enzyme can speed up only one particular reaction. Use these key words in your answer.

| active site | shape | substrate | lock-and-key |
|---|---|---|---|

........................................................................................

........................................................................................

........................................................................................

# Keeping the best conditions for enzymes

## ① Enzymes and temperature

a  Enzymes are affected by temperature and pH.

Write notes in each of the three boxes to explain what is happening at each point on the graph.

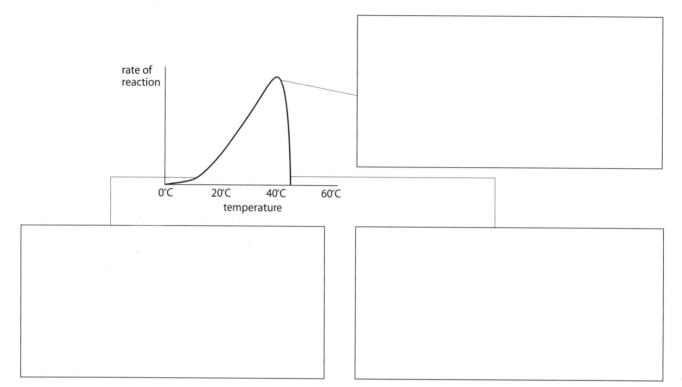

b  The graph shows the effect of pH on an enzyme.

Write down three conclusions that can be made from this data.

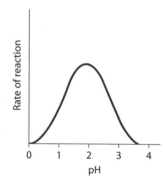

1  ................................................................................................................

2  ................................................................................................................

3  ................................................................................................................

c  Amylase is an enzyme found in saliva in the mouth. The pH of the mouth is about pH 7. The pH of the stomach is about pH 2.

Explain why amylase will not continue working when the saliva is swallowed into the stomach.

................................................................................................................

................................................................................................................

## Enzymes at work in plants

### ① Photosynthesis

Choloroplasts contain the green pigment chlorophyll and the enzymes that are needed for photosynthesis.

**a** Draw a straight line from each chemical to its correct description.

| Chemical | | Description |
|---|---|---|
| carbohydrates | | a type of carbohydrate, synthesised by plants using energy from light |
| chloroplasts | | a type of carbohydrate, used by plants to store energy |
| chlorophyll | | chemicals made of carbon, hydrogen, and oxygen |
| glucose | | plant cell organelles where photosynthesis takes place |
| starch | | the green pigment needed for photosynthesis |

**b** Write down the symbol equation for photosynthesis in the spaces below.

.................................... + .................................... $\longrightarrow$ .................................... + ....................................

**c** Describe the reactions that are taking place during photosynthesis, starting with the absorption of light energy.

**d** Suggest **two** ways that animals depend upon plants for their survival.

### ② Glucose is made during photosynthesis

**a** Describe and explain three different ways in which plants use glucose.

**1**

**2**

**3**

**b** During growth glucose can be converted into other chemicals.

Write down the names of three of these chemicals.

1 ..................................................................................................................................

2 ..................................................................................................................................

3 ..................................................................................................................................

## ① Molecules in gases and liquids move about randomly

**a** Look at the diagram of a cross-section through a leaf.

guard cell

sub-stomatal cavity

Explain how oxygen passes from the cells where photosynthesis takes place to the outside atmosphere.

...........................................................................................................................................

...........................................................................................................................................

...........................................................................................................................................

...........................................................................................................................................

**b** Write down the names of two other chemicals that move in and out of plant cells by diffusion.

1 ...................................................................................... 2 ......................................................

**c** Below is a list of features that adapt the leaf to photosynthesis. Tick ✓ those that help with the diffusion of carbon dioxide and oxygen into and out of the leaf.

stomata (pores) ☐

waxy cuticle ☐

tightly packed rectangular cells in upper surface of leaf ☐

cells packed full of chloroplasts ☐

round spongy cells with air spaces in between ☐

**F** | **Osmosis**

## ① Osmosis – a special case of diffusion

**a** Write down a definition of osmosis.

........................................................................................................................

........................................................................................................................

**b** The diagram shows a partially permeable bag filled with sugar solution. It is sitting in a beaker of a lower concentration of sugar solution.

Draw arrows to show the movement of water molecules between the two solutions.

(Remember, water molecules will move in *both* directions. Use different sized arrows to show the *overall* movement of water molecules.)

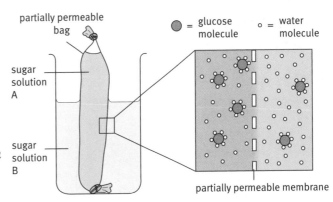

**c** Plant cells can change shape if their water balance is upset.
An onion cell was put into a very dilute sugar solution.

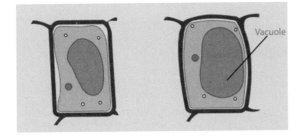

- The solution was more dilute than the contents of the cell.

- There was a higher concentration of dissolved molecules in the cell than in the solution.

- Water entered the cell by osmosis, so the cell swelled up.

A second onion cell was put into a concentrated sugar solution.

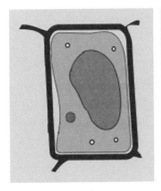 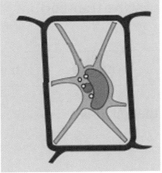

Write three points to explain what has happened to this second cell.

1 .............................................................................................................................

2 .............................................................................................................................

3 .............................................................................................................................

## ② Water moves between different cells by osmosis

Look at the diagram of four cells. Each cell contains glucose but it is at a different concentration in each cell.

**a** Draw a series of arrows to show how water will move by osmosis from cell to cell.

**b** Large carbohydrate molecules such as starch are insoluble. Explain why, for plant cells, starch is a better storage molecule than glucose.

.............................................................................................................................

.............................................................................................................................

## G    Minerals from the soil

## ① Active transport

Active transport is used when a cell needs to take in molecules that are in higher concentration inside the cell than outside the cell.

**a** Complete the sentences to describe diffusion and active transport. Use words from the box.

| active | energy | high | low | passive | transport |
| --- | --- | --- | --- | --- | --- |

Diffusion does not need any ......................... from the cell. It is a ......................... process.

Diffusion only moves molecules from a ......................... to a .........................

concentration. Some molecules move into cells against a concentration

gradient by a process called ..................................................... .

**b** Write down the words in bold below and for each word write down the matching number from the diagram.

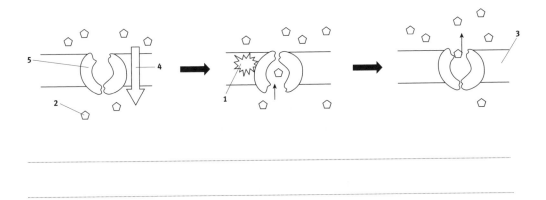

.......................................................................................................................................

.......................................................................................................................................

.......................................................................................................................................

Active transport uses **energy** to move **molecules** across **cell membrane** from low to high concentrations. This is against their **diffusion gradient**. The energy is used to change the shape of a **carrier protein** in the membrane.

## ② Moving nitrate ions

Proteins are long chains of amino acids. To make amino acids, nitrogen must be combined with carbon, hydrogen, and oxygen atoms from glucose.

Use these words to complete the sentences.

| active transport | diffusion | energy | protein |
|---|---|---|---|

The nitrate ions needed for ........................................ synthesis are taken in

from the soil by plant root hairs. To move nitrate ions into the cell, against the

natural ........................................ gradient, uses ........................................ . This is an

example of ........................................ .

## The rate of photosynthesis

## ① Photosynthesis – limiting factors

**a** Explain how the rate of photosynthesis will affect the growth of a plant.

.......................................................................................................................................

.......................................................................................................................................

.......................................................................................................................................

.......................................................................................................................................

**b** Describe three factors that could slow the rate at which photosynthesis takes place.

1 ................................................................................................

2 ................................................................................................

3 ................................................................................................

**c** Carbon dioxide forms 0.04% of air. The graph shows the effect of light intensity on the rate of photosynthesis in normal air at 20 °C. Use the graph to answer the questions.

   **i** What is the main factor limiting photosynthesis at light intensity A?

   ................................................................................................

   **ii** What factors might be limiting photosynthesis at light intensity B?

   ................................................................................................

   **iii** Add and label a second line to the graph to show the effect of an increase in the carbon dioxide concentrationto 0.1%.

   **iv** Add and label a third line to the graph to show the effect of 0.1% carbon dioxide and an increase in temperature to 30 °C.

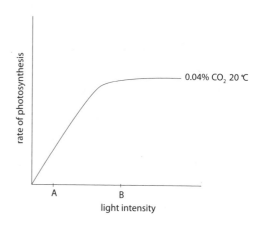

**d** A student collected data about the rate of photosynthesis in pondweed by counting the rate of bubble production under different conditions. Suggest some limitations of the student's data.

................................................................................................

................................................................................................

................................................................................................

................................................................................................

---

## I    Environments and adaptations

### ① Surveying plants

The amount of light available in a habitat affects the types of plant that grow there.

An ancient meadow is under threat because a new road bypass might be built. A conservation group wants to survey the meadow to find out which plants grow there.

**a** The table below lists the equipment they have collected together for the survey. Draw a line from each piece of equipment to its correct description. The first one has been done for you.

|  | |  |
|---|---|---|
| 1 m² quadrat | | to measure light intensity at each quadrat and at 1-m intervals on the transect |
| plastic bags and labelling pen | | to photograph plants for identification, size, and so on |
| identification books and keys | | to form the transect line and measure 1-m intervals |
| tape measure | | to put in any samples that may need to be taken away for identification |
| lightmeter | | to be able to see plant structures more easily to help identification |
| digital camera | | to provide a standard sampling area |
| magnifying lens | | to identify the plants growing in the quadrat |

**b** The conservationists want to take random samples of the meadow. They decide to split the site into nine equal sections. In each section they are going to sample random quadrats. They have thought of two ways they could select the random quadrats:

**Method A**

- Find the dimensions of the sections using paces.
- Use a random-number generator to produce random coordinates.
- Find the random quadrat position by taking the required number of paces.
- Measure the percentage cover of the plants in each random quadrat.

**Method B**

- Blindfold one person and spin them round.
- Get the person to randomly toss the quadrat.
- Measure the percentage cover of the plants in each random quadrat.

    **i** Which method would you choose? ...........................................................................................

    **ii** Suggest three reasons for your choice.

       **1** ...............................................................................................................................

       **2** ...............................................................................................................................

       **3** ...............................................................................................................................

**c** A large oak tree is growing in the middle of the field. The conservationists want to know how the shade from this affects the types of plant growing in the field. Suggest how they can measure this.

......................................................................................................................................................

......................................................................................................................................................

......................................................................................................................................................

---

## J   Energy for life

### ① Respiration

**a** Complete the word equation that sums up the process of aerobic respiration.

....................................     ....................................     ....................................

**b** The rates of photosynthesis and respiration in a plant were measured over 24 hours.

Use the information in the graph to describe how photosynthesis and respiration vary over 24 hours.

......................................................................................................................................................

......................................................................................................................................................

......................................................................................................................................................

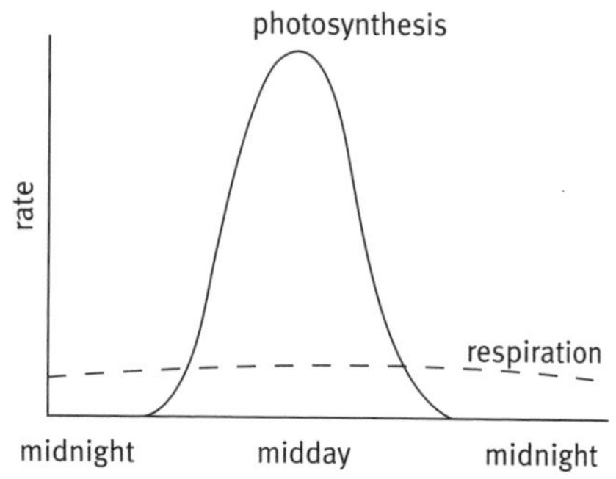

photosynthesis

rate

respiration

midnight      midday      midnight

## ② Respiration in animal cells

**a** The diagram shows the structures used in respiration in animal cells.
Label each structure in the diagram using the boxes on the left.
Then explain what each structure does using the boxes on the right.

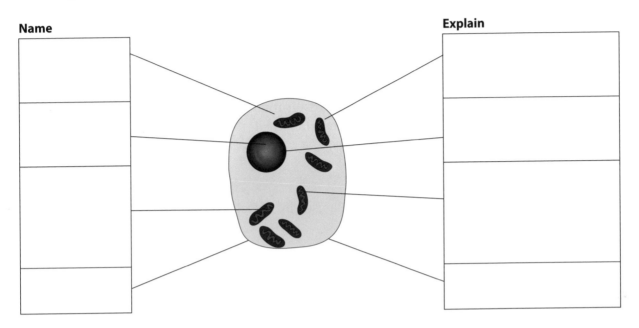

**Name**

**Explain**

**b** Energy from cell respiration is used to make polymers in plants and animals.

  **i** Complete the diagram showing stages in the synthesis of plant cell polymers.

  **ii** Colour the arrows that use energy from respiration.

glucose

............................................. (for storage)

............................................. (for cell walls)

nitrates

amino acids

proteins (for ..................... )

# Anaerobic respiration

**1 a** Energy can be provided by anaerobic respiration.

**i** Write down the word equation for anaerobic respiration in animals in the space below.

..................................................................................................................................................

..................................................................................................................................................

**ii** Write down the word equation for anaerobic respiration in yeast in the space below.

..................................................................................................................................................

..................................................................................................................................................

**b** Explain why anaerobic respiration can only be used by muscles for a short period of time.

..................................................................................................................................................

..................................................................................................................................................

**c** Compare aerobic and anaerobic respiration in human muscle cells by filling in the table.

| | Aerobic respiration | Anaerobic respiration |
|---|---|---|
| What is the energy source? | | |
| What else is needed? | | |
| What waste products are formed? | | |
| Compare the efficiency (which produces the most energy per molecule of glucose). | | |
| Give some examples of when it is useful to the body. | | |
| Describe the effects of increasing this respiration. | | |
| Describe any after-effects of increased respiration. | | |
| Other notes. | | |

**d** Some plants such as mangrove trees live in swamps. Their roots are under water most of the time. The roots contain large cells filled with air. However, most plants can only survive water-logging for a short time. They will die if it continues.

Explain why. ........................................................................................................................

## ② Bacteria

Complete the empty boxes to **name** each structure and **explain** what each structure does.

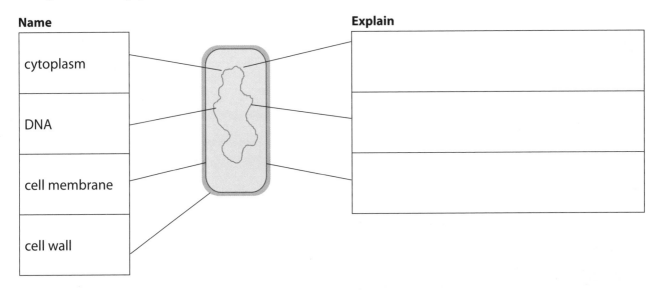

**Name**

| |
|---|
| cytoplasm |
| DNA |
| cell membrane |
| cell wall |

**Explain**

| |
|---|
| |
| |
| |

## Useful products from respiration

## ① Fermentation

Fermentation is a process that happens when yeast cells respire anaerobically. It converts sugars to ethanol and carbon dioxide.

Draw a ring around each by-product of fermentation that is important in producing each of the following products.

Each box may contain more than one ring.

| beer | ethanol / carbon dioxide |
|---|---|
| sparkling wine | ethanol / carbon dioxide |
| bread | ethanol / carbon dioxide |
| biofuel | ethanol / carbon dioxide |
| vinegar | ethanol / carbon dioxide |
| pizza | ethanol / carbon dioxide |

## ② Homemade wine

A student set up this experiment.

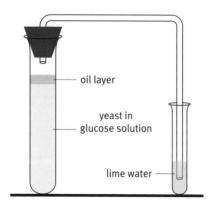

oil layer

yeast in glucose solution

lime water

The student wanted to know how quickly the yeast was respiring.

He counted the number of bubbles given off in one minute, every hour, for eight hours.

These are his results.

| Reading | 1 | 2 | 3 | 4 | 5 | 6 | 7 | 8 | 9 | 10 |
|---|---|---|---|---|---|---|---|---|---|---|
| Number of bubbles | 58 | 56 | 62 | 41 | 59 | 60 | 48 | 36 | 21 | 15 |

a Identify which of the student's results is anomalous and justify your decision in either retaining or discarding the outlier.

...........................................................................................................................................................................

...........................................................................................................................................................................

b What factors should the student control when carrying out the experiment?

...........................................................................................................................................................................

...........................................................................................................................................................................

c Suggest why counting bubbles may not be the best way to determine the rate of respiration.

...........................................................................................................................................................................

...........................................................................................................................................................................

## ③ **Biogas**

Biogas is a fuel produced from manure.

**a** Complete the labels in the diagram below.

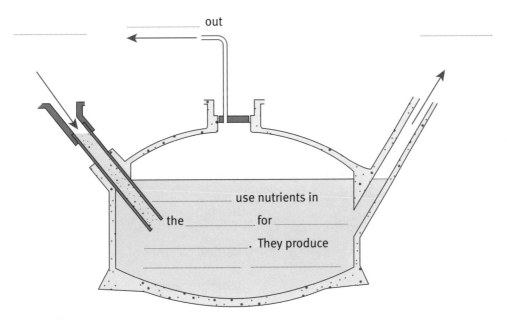

out

use nutrients in

the _____ for _____

. They produce

**b** List three advantages of using biogas rather than using fossil fuels.

1 _____

2 _____

3 _____

**c** Name one disadvantage of using biogas rather than using fossil fuels.

_____

## ① The elements in order

Complete the sentences and diagram.

Scientists look for patterns in data. When they arrange the known elements in order

of relative _____ _____, they find that there

is a repeating pattern. These patterns are shown clearly when the elements are arranged

in a _____. Each row in the table is a called period,

with metals on the _____ and _____ on the right.

G

A column of elements with similar properties →

P                          D

Vertical columns of elements in the

table are _____ of elements

with similar properties.

## ② The periodic table

**a** Colour the key, then use the colours to show these parts on the periodic table below.

☐ Group 1 (alkali metals)  ☐ Group 7 (halogens)  ☐ transition metals

**b** Over three-quarters of the elements are metals. Use another colour to lightly colour in the metals.

| | 1 | | | | | | | | | | | | 3 | 4 | 5 | 6 | 7 | 0 |
|---|---|---|---|---|---|---|---|---|---|---|---|---|---|---|---|---|---|---|
| 1 | 1 H hydrogen 1 | 2 | | | | | | | | | | | | | | | | 4 He helium 2 |
| 2 | 7 Li lithium 3 | 9 Be beryllium 4 | | | | | | | | | | | 11 B boron 5 | 12 C carbon 6 | 14 N nitrogen 7 | 16 O oxygen 8 | 19 F fluorine 9 | 20 Ne neon 10 |
| 3 | 23 Na sodium 11 | 24 Mg magnesium 12 | | | | | | | | | | | 27 Al aluminium 13 | 28 Si silicon 14 | 31 P phosphorus 15 | 32 S sulfur 16 | 35.5 Cl chlorin 17 | 40 Ar argon 18 |
| 4 | 39 K potassium 19 | 40 Ca calcium 20 | 45 Sc scandium 21 | 48 Ti titanium 22 | 51 V vanadium 23 | 52 Cr chromium 24 | 55 Mn manganese 25 | 56 Fe iron 26 | 59 Co cobalt 27 | 59 Ni nickel 28 | 63.5 Cu copper 29 | 65 Zn zinc 30 | 70 Ga gallium 31 | 73 Ge germanium 32 | 75 As arsenic 33 | 79 Se selenium 34 | 80 Br bromine 35 | 84 Kr krypton 36 |
| 5 | 86 Rb rubidium 37 | 88 Sr strontium 38 | 89 Y yttrium 39 | 91 Zr zirconium 40 | 93 Nb niobium 41 | 96 Mo molybdenum 42 | 98 Tc technetium 43 | 101 Ru ruthenium 44 | 103 Rh rhodium 45 | 106 Pd palladium 46 | 108 Ag silver 47 | 112 Cd cadmium 48 | 115 In indium 49 | 119 Sn tin 50 | 122 Sb antimony 51 | 126 Te tellurium 52 | 127 I iodine 53 | 131 Xe xenon 54 |
| 6 | 133 Cs caesium 55 | 137 Ba barium 56 | 139 La lanthanum 57 | 178 Hf hafnium 72 | 181 Ta tantalum 73 | 184 W tungsten 74 | 186 Re rhenium 75 | 190 Os osmium 76 | 192 Ir iridium 77 | 195 Pt platinum 78 | 197 Au gold 79 | 201 Hg mercury 80 | 204 Tl thallium 81 | 207 Pb lead 82 | 209 Bi bismuth 83 | 209 Po polonium 84 | 210 At astatine 85 | 222 Rn radon 86 |
| 7 | 223 Fr francium 87 | 226 Ra radium 88 | 227 Ac actinium 89 | 104 | 105 | 106 | 107 | 108 | 109 | 110 | 111 | 112 | | | | | | |

symbol → 1 H hydrogen 1 ← proton number / name / atomic mass

group number

period number

# The alkali metals

## ① Properties and reactions

**a** Complete the information about Group 1 elements.
The alkali metals:

* are ........................ – you can cut them with a knife

* are ........................ – but only when freshly cut

* quickly ........................ in moist air – they react with water and oxygen

* have low ........................ – some float on water

* react with water to form ........................ and an ........................

    solution of the metal ........................ For example:

    sodium    +    water              →    sodium hydroxide           + hydrogen

    lithium    +    ........................    →    ........................ ........................    + ........................

We call Group 1 the *alkali* metals because they produce alkaline solutions
when they react with water. They also react vigorously with chlorine.
The products are *colourless* crystalline salts called metal chlorides. For example:

    sodium       +       chlorine            →    sodium chloride

    potassium    +    ........................    →    ........................

**b** State and explain the precautions you must take when using Group 1 metals and alkalis.

* Group 1 metals ........................

    ........................

* Alkalis ........................

**c** Look at the information in the table then answer the questions.

| Alkali metal | Proton number | Melting point (°C) | Boiling point (°C) | Density (g/cm³) |
|---|---|---|---|---|
| lithium | 3 | 181 | 1342 | 0.53 |
| sodium | 11 | 98 | 883 | 0.97 |
| potassium | 19 | 63 | 760 | 0.86 |

**i** How does melting point vary with proton number in Group 1?

........................

**ii** How does boiling point vary with proton number in Group 1?

.................................................................................................................................................

**iii** What further information do you need to decide whether or not there is a pattern to the densities of Group 1 metals?

.................................................................................................................................................

**d i** In the table below, describe the reactions of alkali metals with water (right-hand column). Choose phrases from this box.

| | | |
|---|---|---|
| violent reaction | makes sparks | gas catches fire |
| moves around on water | fizzes gently | melts |
| metal thrown off surface | floats | gives off hydrogen |

**ii** Write these words in the left-hand column of the table to show the trend in reactivity down the group.

| | |
|---|---|
| most reactive | least reactive |

| Reactivity | Name of metal | Description of reaction with water |
|---|---|---|
| | lithium | |
| | sodium | |
| | potassium | |

# C   Chemical equations

## ① Names and formulae

Complete the table, with the help of the periodic table on page 20.

| Chemical | Symbols of the element(s) in it | Formula of the chemical |
|---|---|---|
| hydrogen | H | $H_2$ |
| water | ..................... , ..................... | |
| lithium iodide | Li, ..................... | |
| sodium chloride | Na, ..................... | |
| potassium bromide | K, ..................... | |
| lithium hydroxide | Li ..................... , ..................... | |

## ② Balanced equations

**a** Use these words and symbols to complete the sentences.

| g | aq | s | arrow | l | balanced | atoms |
|---|----|---|-------|---|----------|-------|

In a symbol equation, the number of ............................... of each element on each

side of the ............................... in an equation must be equal. We call this a ...............................

equation. We can also show the reactants and products as solid (...............), liquid (...............),

gas (...............), or aqueous solution (...............).

**b** Follow the steps to write a balanced symbol equation for the reaction between potassium and water.

*Step 1:* Describe the reaction in words.

potassium + water → ............................................................... + ...............................................................

*Step 2:* Write down the formulae for the reactants and products.

............................... + ............................... → ............................... + ...............................

*Step 3:* Balance the equation.

............................... + ............................... → ............................... + ...............................

*Step 4:* Add state symbols.

............................... + ............................... → ............................... + ...............................

**c** Follow the steps to write a balanced symbol equation for the reaction between sodium and chlorine.

*Step 1*: Describe the reaction in words.

sodium    +    chlorine → ...............................................................

*Step 2*: Write down the formulae for the reactants and products.

............................... + ............................... → ...............................

*Step 3*: Balance the equation.

............................... + ............................... → ...............................

*Step 4*: Add state symbols.

............................... + ............................... → ...............................

**d** Write a balanced symbol equation for the reaction between potassium and chlorine.

........................................................................................................

## D | The halogens

### ① Properties and reactions

**a** Complete these sentences by drawing a ring around the correct **bold** words.

- The halogens are **metals / non-metals** and their vapours are **coloured / colourless.**

- Halogens can **colour / bleach** vegetable dyes and kill bacteria.

- The halogens are **toxic / non-toxic** to humans.

- Halogen molecules are each made of **one / two** atoms; they are **monatomic / diatomic.**

- Halogens react with **metal / non-metal** elements to form crystalline compounds that are salts.

- The halides of alkali metals are **coloured / colourless** salts such as **potassium / iron** bromide.

- Compounds of Group 1 elements have the formula **MX / MX$_2$** (where M = metal and X = halide).

**b** This is a plot of melting point against proton number for the halogens.

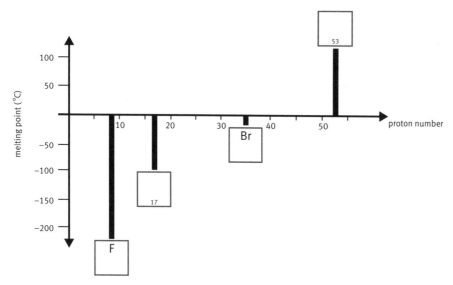

**i** Use the periodic table on page 20 to fill in the two missing proton numbers and two missing symbols.

**ii** Describe the pattern of melting points down the group.

........................................................................................................

**c i** In the table below, describe the reactions of the halogens with hot iron (right-hand column). Write down what you would see during the reaction.

**ii** Write these words in the left hand column of the table to show the trend in reactivity down the group.

| most reactive | least reactive |
| --- | --- |

| Reactivity | Name of halogen | Description of the reaction with hot iron |
| --- | --- | --- |
| | chlorine | |
| | bromine | |
| | iodine | |

## The discovery of helium

## ① Spectra

Spectroscopy has been used in the past to identify new chemical elements.

**a** Colour the lines in this spectrum of helium gas.

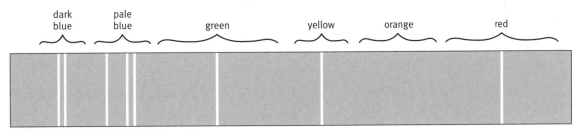

**b** Why is it possible to use spectra to identify elements during chemical analysis?

.................................................................................................................................................

.................................................................................................................................................

**c** How was it possible to discover helium in the Sun before it was discovered on Earth?

.................................................................................................................................................

.................................................................................................................................................

## ② Data and explanations

Below are some data and some explanations for that data.

**a** Decide which of the scientific explanations you accept. Tick ✓ the boxes in the right-hand column of those you think are correct explanations.

| | Data | Explanation | Accept? |
|---|---|---|---|
| 1 | Nitrogen and neon are both unreactive gases. | Nitrogen and neon must be in the same group in the periodic table. | |
| 2 | At room temperature fluorine and chlorine are gases, bromine is a liquid, and iodine is a solid. | The forces of attraction between the molecules become stronger as you go down Group 7. | |
| 3 | Lithium and calcium both react with water to produce hydrogen. | Lithium and calcium must have the same number of electrons in their outer shells. | |
| 4 | A sample of a pure solid produces a line spectrum that has not been seen before. | A new element has been discovered. | |

**b** For each of the explanations, give your reasons for accepting or rejecting it.

1 ...................................................................................................................................

...................................................................................................................................

2 ...................................................................................................................................

...................................................................................................................................

3 ...................................................................................................................................

...................................................................................................................................

4 ...................................................................................................................................

...................................................................................................................................

## Atomic structure

**1** **The parts of an atom**

**a** All atoms are made of the same basic parts – protons, neutrons, and electrons.

Label the diagram of an atom and complete the table.

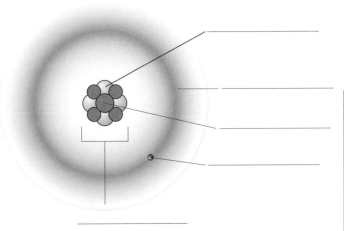

| Part of atom | Relative mass | Charge |
|---|---|---|
| proton | | +1 |
| | 1 | |
| electron | | |

**b** Choose from these words to complete the sentences below.

| electrons | proton | protons | charge | nucleus |
|---|---|---|---|---|

- All atoms of the same element have the same number of ............................................................

- The number of protons is called the ............................................. number.

- The number of protons is equal to the number of ..................................................

  This means that an atom has no overall ..................................................

**2** **Elements and their atomic structures**

Use the periodic table on page 20 to help you complete the table below.

| Element | Proton number | Relative atomic mass | Number of protons | neutrons | electrons |
|---|---|---|---|---|---|
| hydrogen | | | 1 | | |
| carbon | | 12 | | | 6 |
| | 18 | | | | |
| | 3 | | | | |
| | | | 8 | | |

## ① **Electron shells**

**a** Complete these sentences by filling in the blanks with words or numbers.

The electrons in an atom are arranged in a series of _____ around the nucleus.

These shells are also called _____ levels. In an atom the shell closest

to the _____ fills first, then the next shell, and so on.

There is room for:

• up to _____ electrons in shell one.

• up to _____ electrons in shell two.

Once shell two is full, then shell _____ starts to fill.

The _____ number of sodium is 11. So there are _____

electrons in a sodium atom. The diagram below includes the arrangement of _____

in a sodium atom. This electron arrangement can also be written _____.

**b** Show the arrangements of electrons in these atoms. Sodium has been done
for you. You will find the proton numbers in the table on page 20.

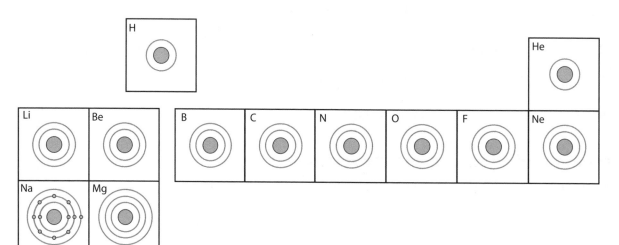

## 1 Electrons and the periodic table

### ① Electron arrangements

**a** Complete these sentences by filling in the blanks with words or numbers.

Shells fill from _____ to _____

across the _____ of the periodic table.

- Shell one fills up first from _____ to helium.

- The second shell fills next from lithium to _____

- Eight _____ go into the third shell from
  sodium to argon.

- Then the fourth shell starts to fill from potassium.

**b** Complete the lists of the arrangement of electrons for these elements:

| Alkali metals (Group 1 in the periodic table) | Halogens (Group 7 in the periodic table) |
|---|---|
| lithium    2.1 | fluorine  _____ |
| sodium  _____ | chlorine  _____ |
| potassium  _____ | bromine  2.8.18.7 |

**c** Since it is the electrons in the outer 'shell' that affect chemical reactions, the number of outer-shell electrons determines the chemical properties of an element.

Complete these sentences.

- Group 1 metals have similar properties because they have one _____ in their outer shell.

- Halogens have _____ properties because they have _____ electrons in their outer shell.

**d** Use the chemical properties of Group 1 and Group 7 elements to illustrate or justify the general statements made in these two paragraphs.

  **i** When atoms react it is the electrons in the outer shell that get involved as chemical bonds break and new chemicals form. Elements have similar properties if they have the same number and arrangements of electrons in the outer shells of their atoms.

_____

_____

29

**ii** The alkali metals are not all the same because their atoms differ in the number of inner full shells. A sodium atom has two inner filled shells, so it is larger than a lithium atom and its outer electron is further away from the nucleus. As a result, the two metals have similar but not identical physical and chemical properties.

................................................................................................................................

................................................................................................................................

## I    Salts

### ① Elements and compounds

An electric current can be used to split the compound lead bromide into its elements.

**a** The diagram below shows the part of the equipment used to split lead bromide into its elements. Label the diagram and complete the sentences by filling in the blanks and putting a ⟨ring⟩ around the correct **bold** words.

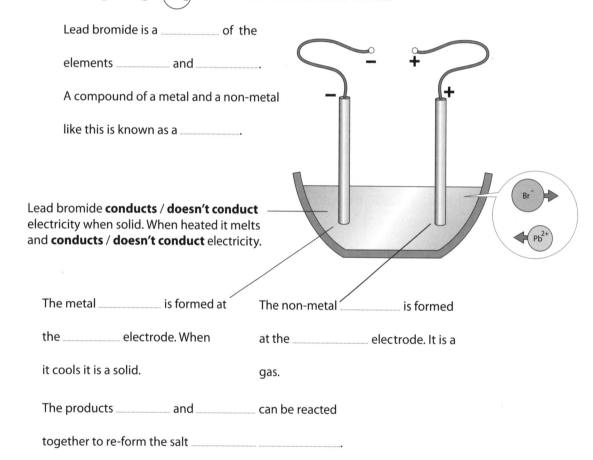

Lead bromide is a ................. of the

elements ............. and ................. .

A compound of a metal and a non-metal

like this is known as a ................. .

Lead bromide **conducts / doesn't conduct** electricity when solid. When heated it melts and **conducts / doesn't conduct** electricity.

The metal ................. is formed at

the ............. electrode. When

it cools it is a solid.

The non-metal ................. is formed

at the ................. electrode. It is a

gas.

The products ............. and ............. can be reacted

together to re-form the salt ................. ................. .

**b** Complete the table to compare the properties of lead bromide, lead, and bromine.

| Property | Lead bromide | Lead | Bromine |
|---|---|---|---|
| state at room temperature | | | |
| colour | white | silver/grey | |
| metal or non-metal | | | |
| element or compound | | | |

## Ionic theory

### ① Ions

Match the beginning and end of each sentence.

| | |
|---|---|
| Charged particles are called . . . | . . . splitting compounds with electricity. |
| Metals form ions that are . . . | . . . negatively charged. |
| Non-metals form ions that are . . . | . . . the positive electrode. |
| Positive ions are attracted to . . . | . . . ions. |
| Negative ions are attracted to . . . | . . . positively charged. |
| Electrolysis is . . . | . . . the negative electrode. |

## ② Salts and ionic theory

**a** The diagram shows a model of the structure of sodium chloride.

   **i** Colour the diagram to show:

- sodium ions *red*

- chloride ions *green*

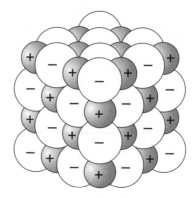

   **ii** How does this model of the structure of sodium chloride explain the shape of sodium chloride crystals?

.............................................................................................................................

.............................................................................................................................

.............................................................................................................................

**b** Complete this table.

| Property of salts | Ionic theory explanation |
|---|---|
| All the crystals of each solid ionic compound are the same shape. Whatever the size of the crystal, the angles between the faces of the crystal are always the same. | |
| | The giant ionic structure is held together by the strong attraction between the positive and negative ions. It takes a lot of energy to break down the regular arrangement of ions. |
| | In a molten ionic compound the positive and negative ions can move around independently. |
| The solution of an ionic compound in water is a good conductor of electricity. | |

# Ionic theory and atomic structure

## ① Atoms into ions

**a** Complete the sentences next to these diagrams.

**i** Atoms of metals on the left-hand side of the periodic table turn into ions by losing electrons.

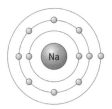

When it turns into an ion

the sodium atom _____

one electron (negative charge)

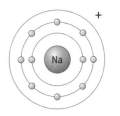

so the sodium ion has a

_____ charge

**ii** Atoms of non-metals to the right of the periodic table turn into ions by gaining electrons.

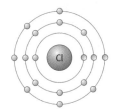

When it turns into an ion

the chlorine atom _____

one electron (negative charge)

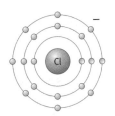

so the chloride ion has a

_____ charge

**b** Complete these diagrams to show the number and arrangement of electrons in each atom and the ions they form. (Your answers to Question 1b in Section G will help you.)

**Atom**

Symbol: Li

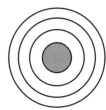

**Ion**

Symbol:  Li⁺

Symbol: Mg

Symbol: _____

Symbol: F

Symbol: _____

Symbol: O

Symbol: _____

## ② Formulae of ionic compounds

This table shows formulae of simple ions.

| | | | | | | | | |
|---|---|---|---|---|---|---|---|---|
| | | $H^+$ | | | | | | |

| | | | | | | | |
|---|---|---|---|---|---|---|---|
| $Li^+$ | | | | | $N^{3-}$ | $O^{2-}$ | $F^-$ |
| $Na^+$ | $Mg^{2+}$ | | $Al^{3+}$ | | | $S^{2-}$ | $Cl^-$ |
| $K^+$ | $Ca^{2+}$ | | | | | | $Br^-$ |
| $Rb^+$ | $Sr^{2+}$ | transition metals form more than one ion, eg $Fe^{2+}$, $Fe^{3+}$ | | | | | $I^-$ |
| $Cs^+$ | $Ba^{2+}$ | | | | | | |
| 1+ | 2+ | | 3+ | | 3− | 2− | 1− |

no simple ions

no ions formed

metals
positive ions

non-metals
negative ions

**a** Complete these general statements about ions.

- The metals in Group 1 and 2 form ................................................................

- The alkali metals form ions with ................................................................

- Non-metals in Groups 6 and 7 form ................................................................

- The halogens form ions with ................................................................

**b** Complete this table to show that ionic compounds are electrically neutral overall because the positive and negative charges balance.

| Ionic compound | Ions present | | Formula |
|---|---|---|---|
| | **Positive ions** | **Negative ions** | |
| | $Li^+$ | | LiBr |
| magnesium iodide | $Mg^{2+}$ | $I^-$ $I^-$ | |
| | | | $AlBr_3$ |
| sodium oxide | | | |

# Chemical species

## ① Chemical hazards

Complete the descriptions of chemical hazards by filling in the blanks.

Then write the names of these chemicals alongside the matching hazards.

Some chemicals have more than one hazard.

| chlorine | bromine | iodine | sodium | sodium hydroxide | dilute hydrochloric acid |

| Symbol | Hazard | Examples |
|---|---|---|
| | **Harmful**<br><br>A chemical that may involve limited health risks if<br><br>breathed in, _____ , or taken in through the<br><br>skin. (Less hazardous than a toxic chemical.) | |
| | **Toxic**<br><br>A chemical that may involve serious health risks<br><br>or even _____ if breathed in, swallowed, or<br><br>taken in through the _____ . | |
| | **Explosive**<br><br>A chemical that may react suddenly and rapidly causing<br><br>an _____ . Explosive chemicals must be<br><br>handled with great care. | The alkali metal caesium reacts explosively with cold water. |
| | **Corrosive**<br><br>A chemical that can destroy living tissue such as<br><br>_____ or _____ . | |
| | **Highly flammable**<br><br>A chemical that may easily catch _____ or<br><br>that gives off a flammable gas in contact with water. | |
| | **Oxidising**<br><br>A chemical that reacts strongly with other chemicals and makes<br><br>the mixture so hot that it may cause a _____ . | potassium nitrate |

## ① Interaction pairs of forces

Forces happen because of an *interaction* between two objects.
They happen in *pairs*.

In each of the bullet points below, draw a ⟨ring⟩ around the correct **bold** word.

The two forces in an interaction pair:

• are **always** / **sometimes** / **never** the same size

• act in **the same** / **random** / **opposite** directions

• act on **the same** / **a different** object

## ② Forces

**a** In each of these drawings, add an arrow in red to show the named force.

the force exerted by Earth on the
apple (the apple's weight)

the force exerted by Tony on the Box

the force exerted by magnet A on magnet B

**b** Complete these sentences by filling in the missing words.
Each sentence will help you with the next one.

**i** Magnet A pulls magnet B to the left; magnet B pulls magnet A

to the _____ with the same size force.

**ii** Tony pushes the box to the right; the box exerts a force on Tony.

It pushes him to the _____ with the _____ size force.

**iii** The Earth exerts a downwards force on an apple; the apple _____

exerts an _____ force on the _____. It pulls back with

the _____ .

**c** Now add a second arrow to each of the pictures above. This should
show the second force in the interaction pair. Use a different colour
and label the force like this:

'force exerted by _____ on _____ '

## Getting moving

### ① Rockets

To start moving, a rocket burns fuel to produce hot exhaust gases.

Explain how the rocket starts moving.

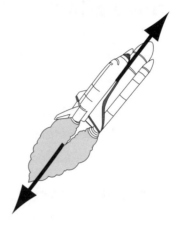

### ② Wheels

A bicycle tyre grips the road. When the wheel turns, the bottom of the tyre pushes backwards on the ground. In the picture, the tyre exerts a force to the left on the ground.

turning wheel

**a** What is the force that allows the tyre to push backwards on the ground?

**b** The ground exerts a force on the tyre. What is the direction of this force?

**c** Draw an arrow on the wheel to show the direction and size of this force.

**d** Label the two arrows showing the forces on the picture.

### ③ Boats

The diagram below shows two boats, with no oars, in the middle of a lake. The boats are close enough for Anna and Sam to push each other.

**a** Draw labelled arrows to show the forces on the diagram, when Anna and Sam are pushing each other.

**b** Explain how this gets the boats moving.

## ① Getting a grip

Look at the picture on the right.
It shows Sophie pulling on a rope
that is attached to a wall.

Draw a circle around two places where
friction is helping her grip.

Sophie

## ② Movement and friction

Use words from the list to complete the sentences below.
Words may be used once, more than once, or not at all.

| upwards | downwards | forwards | backwards | grip | rope | exerts |
|---|---|---|---|---|---|---|

**a**  A racing car accelerates at the start of a race: friction allows the tyres to _____

the track. The tyres exert a force backwards on the track;

the track _____ a force _____ on the tyres.

This makes the car speed up.

**b**  A climber uses a rope to go up a mountain: friction allows her hands to grip the

_____ . Her hands exert a _____ force on the rope;

the rope _____ an _____ force on her hands.

**c**  Mick is trying to walk to the shops. The ground is icy and slippery.

His feet cannot get a _____ to push _____ on the surface.

So the surface does not push him _____ .

## ③ Resultant forces

**a**  Clare and Sophie have a tug-of-war. They pull the rope
with the same amount of force. They are not moving.

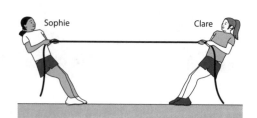

Sophie          Clare

   **i**  Draw and label arrows on the diagram to show:

   • the force exerted by Clare on the rope

   • the force exerted by Sophie on the rope

   **ii**  What is the resultant horizontal force acting on the rope?

_____

**b** A train is made up of 2 engines and 20 carriages.

Each engine provides a force of 30 kN and has friction forces of 2 kN acting on it.
Each carriage has friction forces of 2 kN acting on it.

**i** Fill in the blanks to complete this calculation of the forces on the train.

- total driving force of 2 engines = _____ kN

- total friction forces = friction on _____ engines + friction on _____

  carriages = _____ kN

**ii** Add labelled arrows to the diagram to show the forces acting on the train.

2 engines plus 20 carriages

**iii** Calculate the resultant force on the train.

Resultant force = _____ kN

## Vertical forces

**①** **How surfaces hold things up**

**a** The picture shows a bag on a springy cushion.

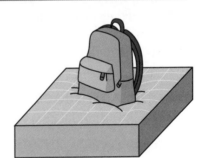

**i** How can you tell that the bag is exerting a force on the cushion?

_____

**ii** Draw a red arrow to show the force that the bag exerts on the
cushion. Label this arrow 'force exerted by bag on cushion'.

**iii** The cushion reacts by pushing up on the bag.
How does the cushion exert a force on the bag?

_____ .

**iv** What is the name of this force? _____ .

**v** Draw a second arrow to show the force the cushion exerts on the bag.
Use a different colour and label this force with its name.

**vi** What would happen to the cushion if the bag were removed?

_____ .

**b** The picture shows an apple resting on a table. Its weight is pulling it down. But the apple is not falling.

   **i** What is the force (and object) that stops the apple from falling?

     ........................................ from the ..............................................

   **ii** Does the table get squashed by the apple? ........................................ .

   **iii** Explain why we don't see the table being squashed.

     ..........................................................................................................

     ..........................................................................................................

   **iv** Explain how the table exerts a force on the apple.

     ..........................................................................................................

     ..........................................................................................................

**c** Complete this sentence to describe the forces between an object and a surface it is resting on.

When an object rests on a surface, it exerts a force .................................................... on the surface.

The surface exerts a force .................................................... on the object.

This is called a .................................................... force.

## ② Weight and reaction

So far, you have looked at pairs of forces acting *between* two different objects.

Now you will work with more than one force acting on a *single* object.

**a** The picture shows a tennis ball on the ground. Its mass is 0.2 kg.

mass = 0.2 kg

   **i** What is its weight? (Take the gravitational field strength as 10 N/kg.)

     weight = ........................ N

     weight = ............ N

   **ii** Write this value on the picture.

   **iii** Draw an arrow to show the reaction force exerted by the ground on the ball.

   **iv** Label this force.

   **v** There are two forces acting on the ball.

     What is the *resultant* force acting on the ball? ........................................

**b** The picture shows a mass hanging on a thread. Its weight is 4 N.

   **i** Draw arrows for the two forces that are acting on the mass.

   Use these phrases to label the following forces:

- the force exerted by the Earth on the mass
- the force exerted by the thread on the mass

  **ii** What is the resultant force on the mass? .............................................................

 **iii** Imagine the thread breaks. Put a cross **×** next to the force that will disappear.

  **iv** Put a tick ✓ next to the force that will still be there after the thread breaks.

   **v** Immediately after the thread breaks, what is the resultant force on the mass?

---

③ **Freefall**

These diagrams show a skydiver falling and opening a parachute. A parachute works by making the air resistance much bigger.

Ⓡ️ⓘⓝⓖ the correct **bold** words.

**a** Air resistance is **less** / **more** than weight.

  Resultant force is **up** / **down** /**zero.**

**b** Air resistance is equal and opposite to weight.

   **i** Draw arrows to show the size and direction of the air resistance and the weight, and label them.

  **ii** Resultant force is **up** /**down** / **zero.**

**c**  **i** Draw arrows to show the size and direction of the air resistance and the weight, and label them.

  **ii** Air resistance is **less** / **more** than weight.
   Resultant force is **up** / **down** / **zero.**

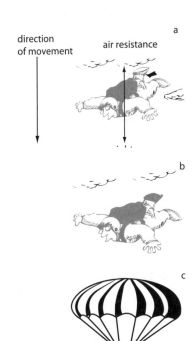

direction of movement   air resistance  a

b

c

## ① Average and instantaneous speed

Where speed is changing, the average speed and instantaneous speed can be different.

The following diagram shows how the positions of six of the runners changed at each 25-metre mark.

| Time: | 0.00 s | 3.00 s | 5.50 s | 8.10 s | 9.97 s |
|---|---|---|---|---|---|
| Distance: | 0 m | 25 m | 50 m | 75 m | 100 m |

**a** Which athlete got the best start to the race (at 3.00 s)? ........................................

**b** Use the information in the diagram to calculate Bailey's average speed over the last 25 m of the race.

Bailey's average speed for the last 25 m = ........................................ m/s

**c** Complete these sentences using words from the box. You can use words once, more than once, or not at all.

| instantaneous | average | lower | higher | speed |
|---|---|---|---|---|

Bailey's speed at any one moment is called his ........................................ speed.

Just after the start, his instantaneous ........................................ is lower than

it is when he gets going. As a result, his ........................................ speed for the last 25 m

is ........................................ than his ........................................ speed for the whole race.

## ② Speed, velocity, and acceleration

**a** Explain the difference between speed and velocity.

........................................................................................................................

**b** The acceleration of a car is given as 0 to 60 mph in 8 s.

Complete this calculation of its acceleration in m/s².

60 mph = 26.67 m/s

$$\text{acceleration} = \frac{\text{change in speed}}{\text{time taken}} \quad \text{(direction doesn't matter in this case)}$$

$$\text{acceleration} = \frac{\text{........................} \text{ m/s} - \text{................} \text{ m/s}}{\text{................} \text{ s}} = \text{........................................ m/s}^2$$

**c** A car accelerates from 13 m/s to 31 m/s. Its acceleration is 0.72 m/s².

**i** (Ring) the equation you would use to calculate the time taken.

time taken = acceleration × change in speed   time taken = $\dfrac{\text{change in speed}}{\text{acceleration}}$   time taken = $\dfrac{\text{acceleration}}{\text{change in speed}}$

**ii** Calculate the time taken.

Time taken = _____

## Picturing motion

### ① Distance – time graphs

**a** Look at the distance–time graph on the right.

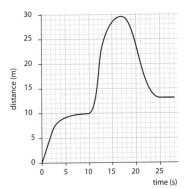

**i** Label the graph with these letters to show where the object is:

  **A** moving away from the start with a constant speed

  **B** moving backwards towards the start with a constant speed

  **C** stationary

**ii** You can tell from the graph that the object moved slower between 0 and 5 seconds than it did between 10 and 15 seconds. Explain how you can see this from the graph.

_____

**iii** Work out the speed of the object between 10 seconds and 15 seconds.

  speed = _____ m/s

**b** Straight slopes on a distance–time graph tell you that an object is moving at a constant speed.

**Curves** tell you that an object is **accelerating**. Its speed is **changing** – it is speeding up or slowing down.

Label the graph with these letters to show where the object is:

**D** slowing down (going forwards)     **E** speeding up (going forwards)

**F** stationary              **G** slowing down (going backwards)

**H** speeding up (going backwards)

## ② Speed–time graphs

A graph of speed against time is another useful way of showing how an object moves.

Look at the distance–time graphs (left) and the speed–time graphs (right) below.
Draw lines to match each distance–time graph to its corresponding speed–time graph.
The first one has been done for you.

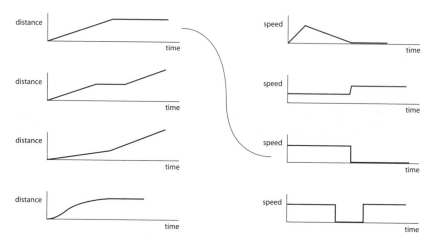

## ③ **Velocity–time graphs**

A tachograph shows the *speed* of a lorry. It is a speed–time graph. You cannot tell which direction it is going.
A *velocity–time* graph shows the *speed* and *direction*.

**a** The velocity–time graphs below show the motion of a car travelling along a straight road running north–south.

Draw lines to match each graph to the best description of the motion. One has been done for you.

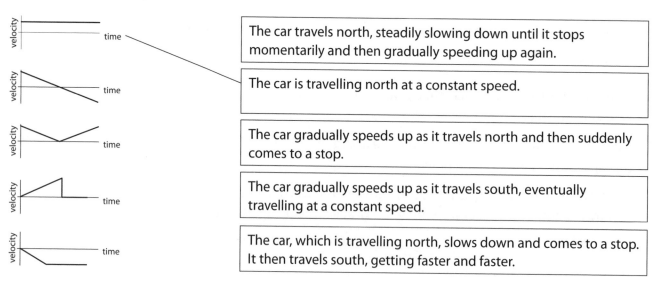

The car travels north, steadily slowing down until it stops momentarily and then gradually speeding up again.

The car is travelling north at a constant speed.

The car gradually speeds up as it travels north and then suddenly comes to a stop.

The car gradually speeds up as it travels south, eventually travelling at a constant speed.

The car, which is travelling north, slows down and comes to a stop. It then travels south, getting faster and faster.

**b** Complete this sentence: When the car is moving backwards, its velocity

is ........................................................................................................ .

## ④ **Acceleration**

Look at the speed–time graph on the right.

**a** Label the graph with these letters to show where the object is:

**A** accelerating so its speed is increasing

**B** accelerating with a negative acceleration so its speed is decreasing

**b** Complete the following statement.

The acceleration of an object can be found from

the ........................ of a speed–time graph.

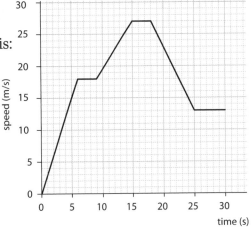

**c** ⟨Ring⟩ the correct calculation for the acceleration between 9 and 15 s and complete the calculation.

$$\frac{(15-9)\,\text{s}}{(27-18)\,\text{m/s}} \qquad \frac{(27-18)\,\text{m/s}}{(15-9)\,\text{s}} \qquad \frac{(27-15)\,\text{m/s}}{(18-9)\,\text{s}} \qquad \frac{(27+18)\,\text{m/s}}{(15+9)\,\text{s}}$$

The acceleration is ........................ $\text{m/s}^2$.

**d** Calculate the acceleration between 0 and 5 s.

acceleration = ........................ $\text{m/s}^2$

## Forces and motion

## ① **Momentum**

| momentum | = | mass | × | velocity |
|---|---|---|---|---|
| (kg m/s) | | (kg) | | (m/s) |

Look at the three balls on the right.

**a** Calculate the momentum of the golf ball.

**b** Use the same method to calculate the momentum of the tennis ball and football.

**c** Each of the balls strikes a large skittle. Put a cross ✗ next to the ball that is most likely to knock it over.

**d** Explain your choice for part **c**.

........................................................................................................................

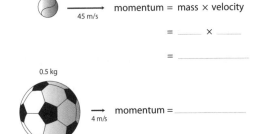

0.05 kg

30 m/s

momentum = mass × velocity

= ......... × .........

= ......... kg m/s

0.05 kg

45 m/s

momentum = mass × velocity

= ......... × .........

= .........

0.5 kg

4 m/s

momentum = ..........

## ② Changing momentum

**a** Jo is ice skating. To get her moving, Sam gives her a push.

> **i** Use words from the list to complete the sentences below.
>
> Words may be used once, more than once, or not at all.

> | zero | momentum | force | left | right | same | opposite |
> |------|----------|-------|------|-------|------|----------|

> Jo is stationary before Sam pushes her. Her momentum is ............... .
>
> Sam exerts a ............... on Jo, pushing her to the ............... .
>
> The force acts for a short time; as a result, her ............... changes.
>
> She gains momentum going to the right – the ............... direction
>
> as the force that made this happen.

> **ii** Sam exerts a force on Jo, which increases her momentum.
>
> Sam could increase Jo's momentum more if he:
>
> • pushes her with more *force* or
>
> • pushes her for a longer *time*
>
> Write in the missing quantities and units to complete the equation:
>
> Change of momentum = ............... × ...............
>
> ...............   ...............   ...............

> **iii** Jo gains 120 kg m/s of momentum when Sam pushes her. Sam exerts a force on her while they are in contact. This lasts for 2 seconds.
>
> What is the size of the average force that Sam exerts on Jo?
>
> Sam's force = ............... N

**b** Lucie and Richard are on their skateboards.

The ground is horizontal and they are not moving.

Lucie has a heavy ball. Lucie throws the ball and Richard catches it.

This sentence describes what happens to Richard and Lucie. Put a ring around the correct **bold** words to complete the sentences.

Richard moves to the **left / right**; Lucie moves to the **left / right**.

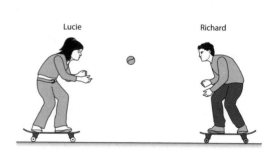

## Car safety

### ① Changing momentum safely

Look at the pictures of two crashed cars. Both cars were travelling at 20 m/s when they hit the wall.

**a** Look at these four phrases. They refer to the force on the driver as he is brought to rest.
Draw lines to link each of the phrases with one of the pictures.

| small time |
| --- |

| big time |
| --- |

| small force |
| --- |

| big force |
| --- |

Car A     0.5 s

Car B     0.05 s

**b** The driver's mass is 60 kg. What is his momentum before the collision?

driver's momentum = _____ kg m/s

**c** He is brought to rest by the force from the seatbelt. The times are shown in the pictures.

Change of momentum = force × time. Calculate the force on the driver in:

  **i** car **A**       **ii** car **B**

**d** It took 10 times longer to stop the driver in car **A**. What effect did this have on the force acting on him?

### ② Seat belt statistics

In 2009 95% of drivers and front-seat passengers and 89% of rear-seat passengers wore seat belts.

About one third of the people who died in cars were not wearing seat belts.

**a** Tick ✓ the true statements, based on the data above.

If 100% of people had worn their seat belt no one would have died. ☐

Wearing a seat belt did not make any difference. ☐

Less than 10% of people travel without wearing a seat belt. ☐

You are much more likely to have a car crash if you are not wearing a seatbelt. ☐

You are more likely to survive a car crash if you are wearing a seat belt. ☐

There is something wrong with the data. If less than 10% have no seat belt, then it's not possible for one third of people who die to be not wearing one. ☐

**b** Complete these statements.

  **i** There is a correlation between wearing a seat belt and _____.

  **ii** The mechanism which suggests that seat belt cause this effect is _____.

## ① First law of motion

> If the resultant force acting on an object is zero, its momentum will stay the same.

It means that steady motion requires no (resultant) force. Complete these sentences. Draw a ⟨ring⟩ around the correct **bold** words.

This means that a **stationary** / **weightless** object will stay where it is.
It **will** / **will not** start moving if there is no resultant force.
A moving object will carry on at **the same** / **a lower** speed.
In both cases the momentum **does** / **does not** change because the resultant force is **zero** / **small**.

## ② Second law of motion

The picture shows Georgia on her bicycle. Imagine Georgia increases her driving force to 150 N.

counter force        driving force

**a** What will be the resultant force be on the bicycle?

_____ N to the _____ .

100 N                    100 N

**b** What will happen to her momentum?

It will _____ to the _____ .

**c** Complete this sentence, which is part of the **second law of motion**.

> When a _____ force acts on an object, its _____ will change in the
>
> _____ direction as the force.

## ③ Third law of motion

Complete this statement of the third law of motion.

> When two objects interact each experiences a _____ _____ . The two
>
> forces are _____ in size but _____ _____ in direction.

## ④ Using Laws 1 and 2

The pictures show forces on some toy cars. Cars A and B are stationary.
The others are moving to the right.

Fill in the blanks for the resultant force and its direction. Then draw lines
to match each of pictures to a correct description of the car's motion.
One has been done for you.

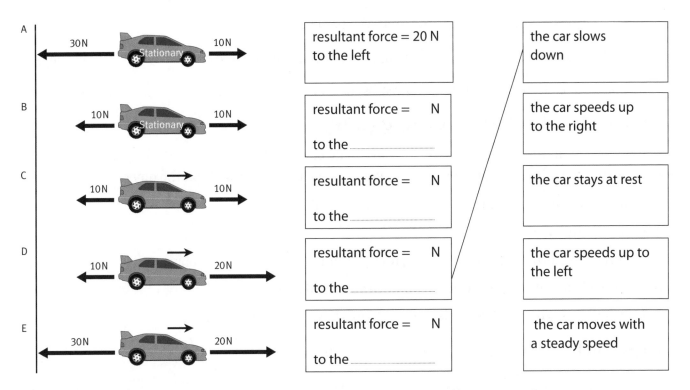

---

## Work and energy

## ① Work

Whenever you move an object, you have to do some work.
The amount of work depends on:

- the force you use
- the distance the object moves

The equation for the amount of work done is:

work done  =  force  ×  distance

(............)      (............)      (............)

**a** Complete the equation by putting the units in the brackets.

**b** Des pushes a box 3 m across the floor. How much work does he do?

work done = _____ N × _____ m

= _____ J

Des

60 N

3 m

**c** Energy is always conserved.

Complete these sentences to describe what happens to the work that Des does.

When Des pushes the box across the floor, the bottom of the box _____ up.

The work goes into _____ the temperature of the box and the ground. When work is done on

an object (or collection of objects), it increases the amount of _____ stored in them.

## ② Gravitational potential energy (GPE)

Whenever you lift an object upwards, you have to do some work.
You are lifting against the force of gravity, which is pulling downwards on the object.

The object then has the potential to do the work back for you.

**a** Complete this sentence.

As a result of lifting the object up, it gains gravitational _____ energy (GPE).

**b** You lift a box that weighs 200 N on to a table that is 0.5 m high.

How much gravitational potential energy (GPE) has the box gained?

gpe gained = _____ N × _____ m

= _____ J

**c** Write down the equation you use to work out change in gravitational potential energy. Include the units.

Change in GPE = _____ × _____

(_____) (_____) (_____)

## ③ Kinetic energy

**a** Complete this sentence.

You can do work on an object to make it speed up.

As a result, the object gains _____

_____ (KE).

Chris

mass of Chris
and bicycle = 110 kg

speed = 8 m/s

distance travelled
speeding up = 50 m

forward force exerted
by Chris = 100 N

**b** Chris does work at the start of a race to get her bike going.
From a standing start, she pedals along a straight road and reaches a speed of 8 m/s.

The mass of Chris with her bike is 110 kg.

**i** Write down the equation you can use to work out *kinetic energy*. Include the units.

Kinetic energy $= \frac{1}{2} \times$ _____ × _____

(_____) (_____) (_____)

**ii** Calculate the kinetic energy of Chris and her bike at 8 m/s.

Kinetic energy =

Kinetic energy of Chris and her bike = ............................... J

**iii** Chris produced a driving force of 100 over the 50 m. Calculate the amount of work she did.

work done = ........................... × ........................... = ........................... J

**iv** Complete these sentences.

Chris did ..................... joules of work; she gained ..................... joules of kinetic energy.

This is ................ than the amount of work she did to get the bicycle moving.

Remember this important principle: *Energy is always conserved.*

**v** Explain what happened to the extra work that Chris did.

................................................................................................................................................

## (4) Roller-coaster energy

**a** Complete these sentences to describe the energy change of a falling object.

When an object falls it loses ...............................

........................... energy but it gains

........................... energy.

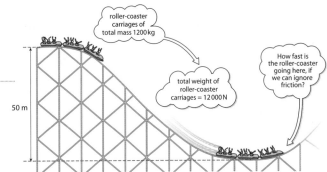

You can use this idea to work out the speed of the rollercoaster carriage at the bottom of the hill.

**b** Write down the equation to work out how much gravitational potential energy the carriage loses and use it to calculate how much gravitational potential energy (GPE) the carriage loses.

GPE lost = ............... J

Energy is always conserved. Assume that all the gpe lost is gained as kinetic energy.

**c** Write down how much kinetic energy the carriage has gained.

kinetic energy gained = ............... J

**d** kinetic energy = $\frac{1}{2}$ mass × (velocity)$^2$

Calculate the size of the carriage's velocity.

carriage's speed = ............... m/s

**e** In reality, how will the speed of the carriage compare with your answer to part **e**? ...............

**f** Explain why.

................................................................................................................................................

**1 a** The following words describe different levels of cell specialisation.

| organ system     tissue     cell     organism     organ |
|---|

Put the words in the correct order.

The first one has been done for you.

**cell** ........................ ........................ ........................ ........................

**b** Each of the words used in part a have a specific meaning.

Draw a straight line form the **word** to its correct **meaning**.

| organ system | | The smallest living structure found in a living creature. |
|---|---|---|
| tissue | | A group of specialised cells. |
| cell | | A group of different types of tissue that work together. |
| organism | | A group of organs that work together. |
| organ | | A living creature. |

**c** Plant cells also contain tissues and organs.

Draw a straight line from each **structure** to show whether it is a **tissue** or an **organ**.

**structure**

| organ | | flower | tissue |
|---|---|---|---|
| | | leaf | |
| | | phloem | |
| | | root | |
| | | stem | |
| | | xylem | |

**d** Complete the following sentences.

Choose the correct word from this list.

| | | | | |
|---|---|---|---|---|
| an egg cell | a zygote | an embryo | identical | meiosis |
| meristems | mitosis | 4 | 8 | 16 | specialised |

A fertilised egg cell is called _____ .

It divides by _____ to form _____ .

In a human embryo, up to the _____ cell stage, all the cells
are _____ . After this stage, cells become _____

## ② Plants can continue growing all their lives

Use these words to complete the sentences. Each word may be used more than once.

| | | | | | | |
|---|---|---|---|---|---|---|
| longer | meristem | roots | shoots | stems | taller | thicker |

Plants have unspecialised cells called _____ cells.  These cells mean that

plants go on growing at the tips of the _____ and _____ , and increase

the width of the _____ .  These unspecialised cells produce different

tissues so that plants continue to grow throughout their life.

## ③ Cell specialisation in human embryos

Cells in an early human embryo can develop into any sort of cell, but soon
cells become specialised to form a particular type of tissue.

**a** The diagram shows a human egg being fertilised and starting to develop.
Choose words from this box to complete the labels.

| | | | | |
|---|---|---|---|---|
| egg cell | embryo | fertilisation | sperm cell | zygote |

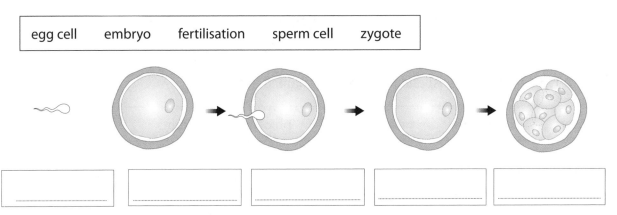

| | | | | |
|---|---|---|---|---|
| ........................ | ........................ | ........................ | ........................ | ........................ |

**b** Up to the eight-cell stage, the cells in the developing embryo are not specialised.

Complete the notes below the diagram to describe an important characteristic of these embryonic stem cells.

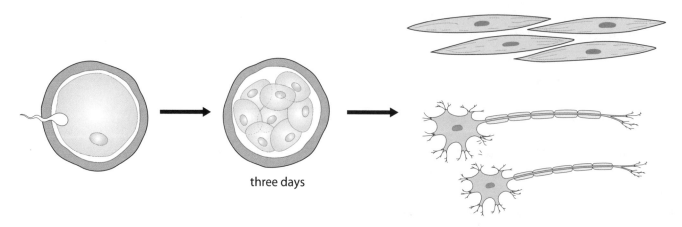

three days

Embryonic stem cells can develop into any type of specialised cell such as ........................

.................................................................................................................................

## ① **Making more plants**

The hormonal conditions in a plant's environment can be changed.
Then unspecialised plant cells can develop into a range of other tissues
or organs.

**a** As a plant grows, cells specialise into tissues. These arrange themselves into organs.

   **i** List the examples of plant tissue given in the box below.

   .................................................................................................................................

   **ii** List the examples of plant organs given in the box below.

   .................................................................................................................................

   **iii** Draw a ring around the *unspecialised plant cell* named in the list.

| flowers | leaves | meristem | phloem | roots | xylem |
|---------|--------|----------|--------|-------|-------|

**b** Use these words to complete the sentences.

| clone | hormones | meristem | roots | tissue |
|-------|----------|----------|-------|--------|

Unspecialised plant cells can make any kind of ........................ the plant needs.

Rooting powder can be used to encourage cut shoots to form ........................ .

Rooting powder contains plant ........................ . These cause the new cells produced by the ............................ cells in the shoot to develop into roots. The

cutting then grows into a complete plant that is a ............................ of the parent.

**c** There are advantages to taking cuttings. One advantage is that you can reproduce plants that are identical to the parent plant. Explain why this is an advantage.

.................................................................................................................................................

**(2)** Unlike humans, plants can only grow in size in special places within the plant.

**a** What are these special growing regions called?

.................................................................................................................................................

**b** Describe three places within plants where these growing regions are found.

**1** ............................................................................................................................................

**2** ............................................................................................................................................

**3** ............................................................................................................................................

**c** Explain the role of meristems when a gardener takes a cutting and grows it into a new plant.

.................................................................................................................................................

.................................................................................................................................................

.................................................................................................................................................

**d** Complete the following sentences about meristems.

Choose the correct words from this list.

| cell | meiosis | mitosis | nucleus | specialised | unspecialised |
|------|---------|---------|---------|-------------|---------------|

Meristem cells divide by ............................ .

The cells are ............................ and can develop into any kind of plant

............................ .

# Phototropism

**(1)** **Phototropism**

**a** Explain **how** phototropism increases a plant's chances of survival.

.................................................................................................................................................

.................................................................................................................................................

.................................................................................................................................................

**b** Name the plant hormone that is involved in phototropism.

.................................................................................................

**c** Rowan is studying phototropism at school.

He thinks that plant growth is controlled by both genes and the environment.

Is Rowan correct?

Explain your answer.

.................................................................................................

.................................................................................................

.................................................................................................

**d** In 1913 scientists carried out three experiments to find out how plants grew towards the light.

In **experiment 1** they cut off the tip of the shoot.

The result was that the shoot did not bend towards the light.

In **experiment 2** they cut off the tip and inserted a piece of impermeable metal.

The shoot did not bend towards the light.

In **experiment 3** they cut off the tip and inserted some permeable jelly.

The shoot bent towards the light.

Which of these conclusions can the scientists make from the experiment?
Put ticks (✓) in the boxes next to the correct answers.

| | |
|---|---|
| Shoots will also bend in the dark when the light is switched off. | |
| Plant shoots grow towards the light. | |
| All plants behave in a similar way. | |
| The response is triggered by the tip of the shoot. | |
| A substance produced in the tip passes down the shoot to trigger the response. | |
| Roots respond by growing away from the light. | |
| The response is caused by a neuron. | |

**e** Describe the effect of light on the distribution of the plant hormone in the shoot tip and how this causes the shoot to grow towards the light.

## A look inside the nucleus

### 1 Chromosomes

The chromosomes in the cell nucleus are made up of many genes.
Each gene is a length of DNA.
Use these words to complete the diagram.

| chromosomes | genes | nucleus |

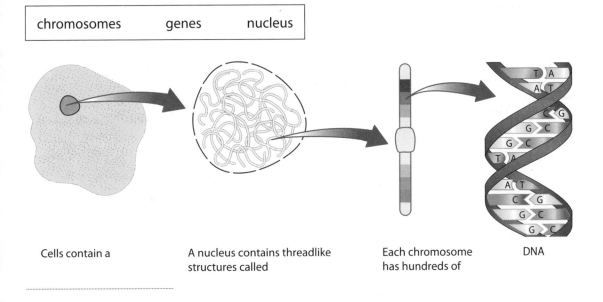

Cells contain a

A nucleus contains threadlike structures called

Each chromosome has hundreds of

DNA

### 2 Blood cells

Mature human red bloods cells are very unusual in that they do not have a nucleus like other cells in the body.

**a** Suggest one advantage to human red blood cells not having a nucleus.

**b** Suggest what effect not having a nucleus will have on red blood cells.

**c** Suggest whether the cells that produce new red blood cells have a nucleus.

Explain your answer.

.................................................................................................................................................

.................................................................................................................................................

## ① **Mitosis**

**a** Living organisms grow by making new cells. The diagram shows the stages in the cell cycle in a cell with four chromosomes (in two pairs). Complete the diagram by writing in the correct words and numbers and drawing the correct diagram.

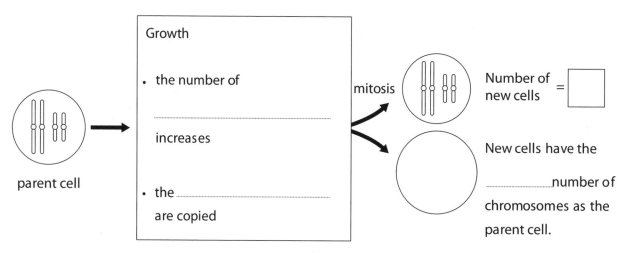

Growth

• the number of

......................................................................

increases

• the ..............................................

are copied

parent cell

mitosis

Number of
new cells = ☐

New cells have the

...........................number of

chromosomes as the

parent cell.

**b** Complete these sentences.

Cell division for growth is called ............................................. For each division, there

are ............................................. new cells produced, which are genetically ...........................................

to the parent cell.

## ② Read these notes form a scientist's journal.

1. It's amazing. When I looked through the microscope for the first time I saw that the plant tissue was made up of little boxes.
2. I think that tissue from other plants may also be made up of little boxes.
3. I have an idea that the little boxes are the smallest units of living tissue.
4. I suddenly realised that some of the boxes are different. I suspect that the different types of box are specialised to do different jobs in the plant.

**a** What do today's scientists call these little 'boxes'

.................................................................................................................................................

**b** Which statements, **1, 2, 3, or 4**, involve data?

.................................................................................................................................................

**c** Which statements, **1, 2, 3, or 4**, involve a prediction?

.................................................................................................................................................

**d** Which statements, **1, 2, 3, or 4**, involve a hypothesis without creative thinking?

.................................................................................................................................................

**e** Which statements, **1, 2, 3, or 4**, involve a hypothesis with creative thinking?

.................................................................................................................................................

**3** The cell cycle of mitosis involves two stages: cell growth and cell division.

Draw a straight line from each **part of the cell cycle** to its correct stage of **growth** or **division**.

**Part of cell cycle**

| | |
|---|---|
| **growth** | number of cell organelles increases |
| | chromosomes are copied |
| | copies of chromosomes separate |
| | the nucleus divides |

**division**

# Sexual reproduction

## ① Meiosis

Cells divide by meiosis when forming gametes (sex cells). This produces four new cells with half the number of chromosomes of the parent cell.

**a** Cell division that produces gametes is called meiosis. Complete the diagram showing meiosis in a cell with four chromosomes (in two pairs) by writing in the correct words, numbers, and diagrams.

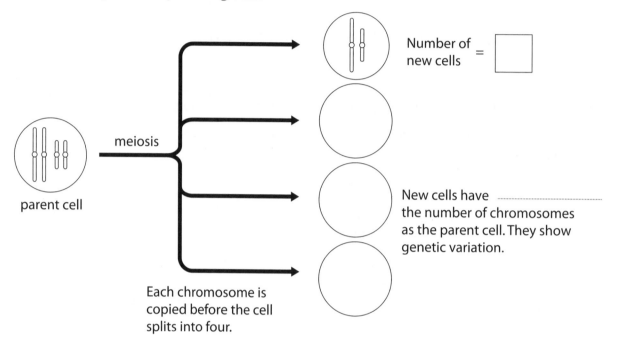

Number of new cells = ☐

New cells have ............................ the number of chromosomes as the parent cell. They show genetic variation.

parent cell

meiosis

Each chromosome is copied before the cell splits into four.

**b** Complete these sentences.

Cell division to form gametes is called ........................................... . There are ...........................................

new cells produced, with half the number of chromosomes of the parent cell.

They show genetic ........................................... .

**c** In human sexual reproduction, a male gamete fuses with a female gamete. The gametes each have half the number of chromosomes of the parent organism. The zygote has a set of chromosomes from each parent.

Fill in the number of chromosomes in each human gamete.

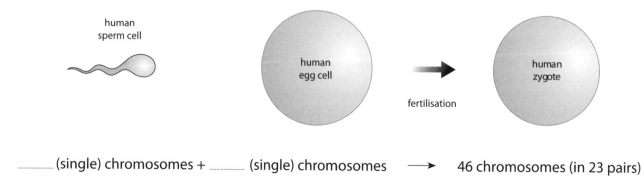

human sperm cell

human egg cell

fertilisation

human zygote

............ (single) chromosomes + ............ (single) chromosomes ⟶ 46 chromosomes (in 23 pairs)

**d** Describe the differences between the two types of cell division, mitosis and meiosis.

.......................................................................................................................

.......................................................................................................................

.......................................................................................................................

.......................................................................................................................

.......................................................................................................................

.......................................................................................................................

**e** Why is it important that gametes have half the number of chromosomes of body cells?

.......................................................................................................................

.......................................................................................................................

## The mystery of inheritance

## ① Copying DNA

When chromosomes are copied, the two strands of each DNA molecule separate.
New strands form alongside the old strands.

**a** Look at the diagram below showing how a double strand of DNA is copied
to form two identical strands.

Complete these sentences.

- There are ............................................ different bases in DNA.

- Base A always pairs with base ............................................ .

- Base C always pairs with base ............................................ .

**b** Complete the sentences under each diagram to show how two exact copies of the DNA are made.

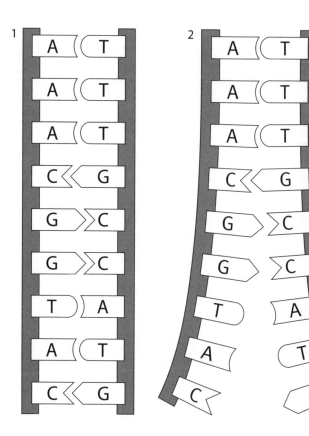

original    new         new    original

........................................
along the DNA are
joined in matching
base pairs.

Weak bonds between the
bases split. The DNA opens

into ..........................................
strands. Free bases in the cell
pair with the bases on each
open strand.

The result is two DNA molecules.
Each molecule is half new. The

.......................................... pairs are in the
same order as in the original DNA.

**c** When DNA was discovered, scientists developed two theories about how DNA was copied.

One theory stated that the whole double helix was copied. This was called the **conservative theory**.

The second theory stated that half of the double helix was copied and replaced by a new strand. This was called the **semi-conservative theory**.

To prove which theory was correct, scientists replaced the normal nitrogen ($N^{14}$) in DNA with a heavy isotope of nitrogen called $N^{15}$.

The diagram shows what they found.

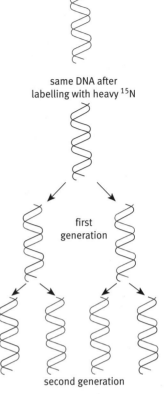

original DNA containing normal $^{14}$N

same DNA after labelling with heavy $^{15}$N

first generation

second generation

**i** What percentage of DNA in the second generation was original DNA?

......................................................................................................................

**ii** What conclusion can you draw from this data?

......................................................................................................................

**iii** Which of the following scientific processes are used in the above example?

Put ticks (✓) in the boxes next to the correct answers.

| | |
|---|---|
| hypothesis | |
| peer review | |
| reproducibility | |
| risk | |
| testing | |
| cause and effect | |

## ① Protein production

During protein production a copy of the gene is made in the nucleus.
This copy carries the DNA instructions to the cytoplasm, where the
protein is assembled.

**a** The diagram shows the steps necessary for an active gene to make a protein.
Carefully label the diagram using these words.

amino acids      copy of the gene      cytoplasm      DNA      nucleus      protein
protein production (in ribosome)

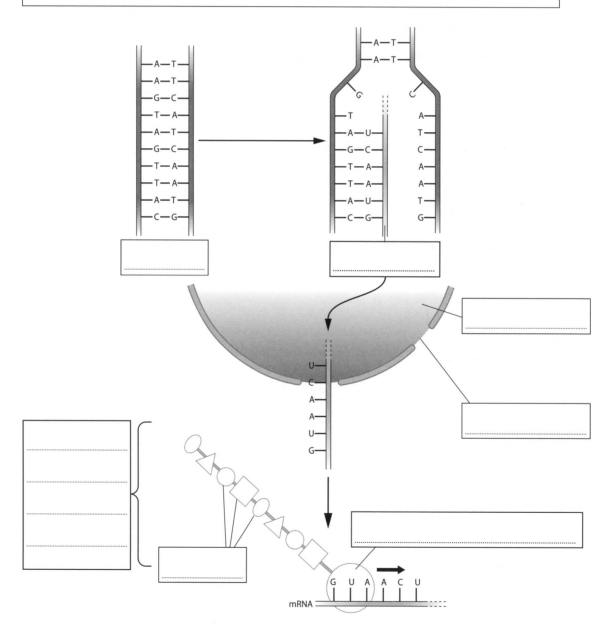

**b** Explain how the order of bases in a gene is the code for making proteins.

## Specialised cells – special proteins

### ① Genes and proteins

Body cells in an organism all contain the same genes. But many genes in a particular cell are not active because the cell produces only the proteins that it needs.

**a** Each gene carries the instructions for a different protein. Proteins have many different functions.

Draw a straight line from each protein substance to its correct description.

| Substance | Description |
|---|---|
| amylase | a structural protein of hair and nails, hard |
| chlorophyll | a structural protein of ligaments, strong |
| collagen | a structural protein of skin, stretchy |
| elastin | a digestive enzyme |
| insulin | a green pigment that absorbs light energy |
| keratin | a hormone made in the pancreas to control blood sugar |

**b** Which of these statements about genes and proteins are correct?

Put ticks (✓) in the boxes next to the correct answers.

| | |
|---|---|
| One gene can produce several different proteins. | |
| All body cells in an organism contain the same genes. | |
| Some genes in a cell are switched off. | |
| For a cell to survive, all its genes must be switched on. | |
| Cells only produce the specific proteins that they need. | |
| Any cell in the body can produce any protein. | |

## ① **Using stem cells**

Stem cells have the potential to produce any type of cell. Stem cells can replace damaged tissue.

**a** The diagram below shows how embryonic stem cells are made. They could be used to make different types of tissue for medical treatment.

Write notes in the boxes to explain the process.

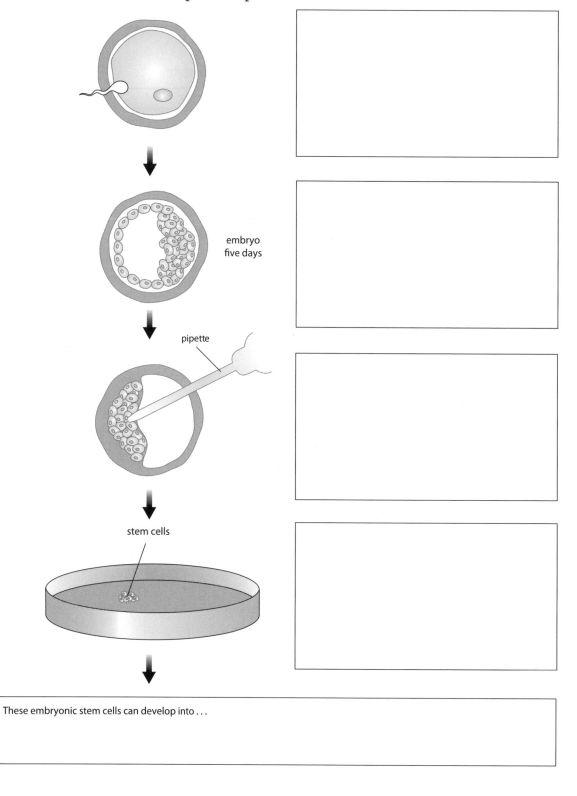

embryo
five days

pipette

stem cells

These embryonic stem cells can develop into . . .

**b** Some people agree with carrying out research on embryonic stem cells. Others do not.

> Stem cell research can be used to treat some serious diseases.

**Mary**

> A few embryos may be destroyed, but think of all the thousands of ill people who could be cured.

**Jake**

> I think it is morally wrong to use cells from an embryo.

**Mike**

> I do not know enough about stem cell research to make an informed decision either way.

**Karen**

**i** Which of the people above is using ethics in their argument?

.................................................................................................................................

**ii** Who thinks that the right decision is the one that leads to the best outcome for the greatest number of people involved?

.................................................................................................................................

**c** Explain the difference between embryonic stem cells and adult stem cells.

.................................................................................................................................

.................................................................................................................................

**d** Suggest why using adult stem cells will involve fewer ethical issues.

.................................................................................................................................

**e** Explain what prevents scientists performing unethical experiments on embryonic stem cells.

.................................................................................................................................

.................................................................................................................................

## Chemicals in spheres

### ① Chemicals in three spheres

Write the names of these chemicals in the boxes on the diagram.
Some chemicals belong in more than one box.

| argon | crude oil | nitrogen | sodium chloride | chalk | iron ore | oxygen |
| water | carbon dioxide | granite | sandstone |

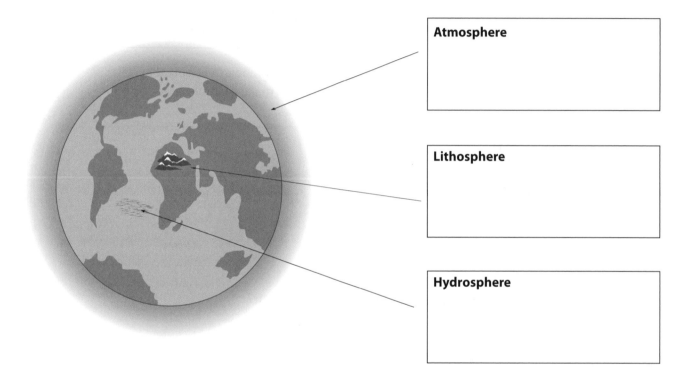

**Atmosphere**

**Lithosphere**

**Hydrosphere**

## Chemicals of the atmosphere

### ① Gases in air

Complete this table to show the gases in unpolluted, dry air.

| Gas | Element or compound? | Percentage by volume in dry air |
|---|---|---|
| | | 78 |
| | element | |
| argon | | 1 |
| | compound | 0.04 |

## ② Strong bonds in molecules

There are strong bonds between atoms in some molecules. These are covalent bonds.
Colour the models and complete the table below to show the bonding in molecules.
Use this information to help you.

| Atom | Usual number of covalent bonds | Colour code in models |
|---|---|---|
| H, hydrogen | 1 | white |
| C, carbon | 4 | black |
| O, oxygen | 2 | red |
| N, nitrogen | 3 | blue |
| Cl, chlorine | 1 | green |

| Chemical | Molecular model | Diagram showing covalent bonds in the molecule | Molecular formula |
|---|---|---|---|
| hydrogen | | H—H | $H_2$ |
| | | | |
| nitrogen | | | |
| water | | | |
| | | | |
| methane | | | |
| chlorine | | | |
| ammonia | | | |
| ethene | | | |

### ③ Covalent bonds

The diagram represents how a covalent bond holds two hydrogen atoms together to make an $H_2$ molecule.

Use a different colour to colour in each phrase/word printed in **outline** in the sentence below.

Then shade the corresponding part of the diagram with the same colour.

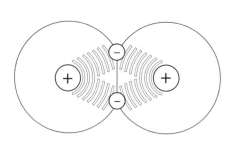

In the hydrogen molecule ($H_2$) the two hydrogen atoms are held together by the **electrostatic attraction** between each of the **nuclei** and the **shared pair of electrons**.

---

| C | Chemicals of the hydrosphere |
|---|---|

### ① Water explanations

Explain each of the following statements.

**a** Chemicals made of small molecules usually have low boiling points.

.........................................................................................................................

.........................................................................................................................

.........................................................................................................................

**b** The boiling point of water is higher than other molecules of a similar size.

.........................................................................................................................

.........................................................................................................................

.........................................................................................................................

**c** Pure water does not conduct electricity.

.........................................................................................................................

.........................................................................................................................

.........................................................................................................................

**d** You should not touch electrical devices with wet hands.

.........................................................................................................................

.........................................................................................................................

.........................................................................................................................

② # Data for the hydrosphere

Look at the tables and diagram below which give information about the hydrosphere.

| Element | Seawater composition (% by mass) |
|---|---|
| oxygen | 85.84 |
| hydrogen | 10.82 |
| chlorine | 1.94 |
| sodium | 1.08 |
| magnesium | 0.1292 |
| sulfur | 0.091 |
| calcium | 0.04 |
| potassium | 0.04 |
| bromine | 0.0067 |
| carbon | 0.0028 |

①

②

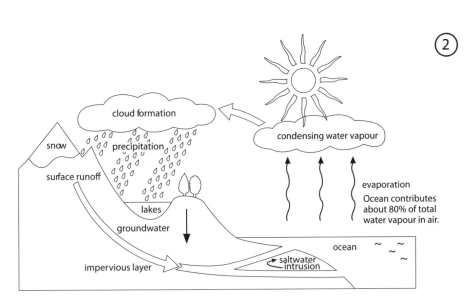

| Gas | Formula | % in atmosphere | % in surface seawater | ml/litre seawater | mg/kg (ppm) in seawater |
|---|---|---|---|---|---|
| nitrogen | $N_2$ | 78 | 47.5 | 10 | 12.5 |
| oxygen | $O_2$ | 21 | 36.0 | 5 | 7 |
| carbon dioxide | $CO_2$ | 0.04 | 15.1 | 40 | 90 |
| argon | Ar | 1 | 1.4 | 0 | 0.4 |

③

For each question write down the number of the table or diagram that can be used to answer it: 1, 2 or 3. If you cannot answer a question using the data given, write X.

**a** Where does most of the water vapour in the air come from? ............

**b** Which element is present in the highest concentration in seawater? ............

**c** What is the percentage of fluorine in seawater? ............

**d** Which gas in the atmosphere is most soluble in seawater? ............

**e** Which metallic element has the highest concentration in seawater? ............

**f** Is there more oxygen or nitrogen dissolved in seawater? ............

**g** Why is there much less bromine than chlorine in seawater? ............

## ① Ionic equations

**a** When calcium nitrate and sodium carbonate are mixed together, they react to form a precipitate of calcium carbonate.

This can be shown as a balanced equation:

$Ca(NO_3)_2(aq) + Na_2CO_3(aq) \rightarrow CaCO_3(s) + 2NaNO_3(aq)$

or as an ionic equation:

$Ca^{2+}(aq) + CO_3^{2-}(aq) \rightarrow CaCO_3(s)$

**i** What does the symbol (aq) mean? ................................................

**ii** What does the symbol (s) mean? ................................................

**b** Write a balanced equation and an ionic equation for each of the reactions below. Use the table of ions to help you.

| $Na^+$ | $Ca^{2+}$ | $Li^+$ | $Ba^{2+}$ |
|---|---|---|---|
| $Al^{3+}$ | $K^+$ | $CO_3^{2-}$ | $NO_3^-$ |

**i** Potassium carbonate is mixed with calcium nitrate to form a precipitate of calcium carbonate.

......................................................................................

......................................................................................

**ii** Barium nitrate is mixed with sodium carbonate to form a precipitate of barium carbonate.

......................................................................................

......................................................................................

**iii** Lithium carbonate is mixed with aluminium nitrate to form a precipitate of aluminium carbonate.

......................................................................................

......................................................................................

## ② Making precipitates

This table gives the solubility of some salts and hydroxides.

| Salt or hydroxide | Solubility |
|---|---|
| all nitrates | soluble |
| all salts of sodium and potassium | soluble |
| silver iodide | insoluble |
| barium sulfate | insoluble |
| calcium carbonate | insoluble |
| hydroxides of metals not in Group 1 | insoluble |

An experimenter mixed some solutions together. Put a tick ✓ in each box where you would expect them to see a precipitate, and a cross ✗ in each box where you would not expect them to see a precipitate.

| Solutions mixed | Precipitate forms? |
|---|---|
| sodium nitrate and calcium chloride | |
| aluminium nitrate and sodium hydroxide | |
| silver nitrate and sodium iodide | |
| potassium bromide and sodium carbonate | |
| lithium hydroxide and calcium nitrate | |
| copper nitrate and sodium hydroxide | |

## ③ Analysing salts

This table shows the results of tests for positively charged ions.

| Metal ion tested | Test | Observation |
|---|---|---|
| calcium $Ca^{2+}(aq)$ | add sodium hydroxide solution | white precipitate (insoluble in excess) |
| copper $Cu^{2+}(aq)$ | add sodium hydroxide solution | light-blue precipitate (insoluble in excess) |
| iron(II) $Fe^{2+}(aq)$ | add sodium hydroxide solution | green precipitate (insoluble in excess) |
| iron(III) $Fe^{3+}(aq)$ | add sodium hydroxide solution | red–brown precipitate (insoluble in excess) |
| zinc $Zn^{2+}(aq)$ | add sodium hydroxide solution | white (dissolves in excess hydroxide solution) |

This table shows the results of tests for negatively charged ions.

| Ion tested | Test | Observation |
|---|---|---|
| carbonate $CO_3^{2-}$(aq) | add dilute acid | effervesces, and carbon dioxide gas produced (the gas turns limewater milky) |
| chloride $Cl^-$(aq) | acidify with dilute nitric acid, then add silver nitrate solution | white precipitate, AgCl(s) |
| bromide $Br^-$(aq) | acidify with dilute nitric acid, then add silver nitrate solution | cream precipitate, AgBr(s) |
| iodide $I^-$(aq) | acidify with dilute nitric acid, then add silver nitrate solution | yellow precipitate, AgI(s) |
| sulfate $SO_4^{2-}$(aq) | acidify, then add barium chloride solution or barium nitrate solution | white precipitate, $BaSO_4$(s) |

a Solution V gives a light-blue precipitate when sodium hydroxide solution is added and a white precipitate when acidified barium nitrate solution is added.

What salt does solution V contain? ............................................................

b Solution W gives a white precipitate when sodium hydroxide solution is added. The precipitate does not dissolve in excess sodium hydroxide. When some of solution W is acidified and silver nitrate is added a yellow precipitate forms.

What salt does solution W contain? ............................................................

c You have three bottles of solutions but the labels have fallen off. You know that one contains sodium chloride solution, another contains zinc chloride solution, and another contains iron(II) chloride solution.

What test(s) would you carry out to find out which is which, and what results would you expect to see?

............................................................................................................................

............................................................................................................................

............................................................................................................................

............................................................................................................................

d Write ionic equations for the following reactions:

i silver nitrate reacting with potassium bromide ............................................................

ii sodium hydroxide reacting with calcium chloride ............................................................

iii barium nitrate reacting with lithium sulfate ............................................................

# Chemicals of the lithosphere

## ① Properties of ionic compounds

The properties of ionic compounds are explained by their structure.

**a** Colour the diagram of the structure of sodium chloride. Colour the sodium ions *red* and the chloride ions *green*. Then complete the labels.

Sodium ions form when sodium atoms ............................ electrons. Each sodium atom loses one ............................ to turn into an ion.

Chloride ions form when chlorine atoms ............................ electrons. Each chlorine atom gains ............................ electron to turn into an ion.

Opposite charges ............................ . So the positive sodium ions strongly attract the negative chloride ions. This is ............................ bonding.

This is a small part of a ............................ ............................ structure. A crystal of sodium chloride consists of millions and millions of ............................ .

**b** Draw lines to match each property to the best explanation for the property.

| Property | Explanation |
| --- | --- |
| All the crystals of each solid ionic compound are the same shape. Whatever the size of the crystal, the angles between the faces of the crystal are always the same. | The giant ionic structure is held together by the strong attraction between the positive and negative ions. It takes a lot of energy to break down the regular arrangement of ions. |
| The solution of an ionic compound in water is a good conductor of electricity. | The ions in the giant ionic structure of an ionic compound are always arranged in the same regular way. |
| Ionic compounds have relatively high melting points. | In a molten ionic compound the positive and negative ions can move around independently. |
| When an ionic compound is heated above its melting point, the molten compound is a good conductor of electricity. | In a solution of an ionic compound, the positive metal ions and the negative non-metal ions can move around independently. |

## ① Diamond and graphite

Diamond and graphite are both made of carbon, but the atoms are held together in different ways. Diamond has the following properties: high melting and boiling points, very hard, insoluble, and does not conduct electricity.

**a** Which two of these properties does graphite *not* share with diamond?

1 ........................................................... 2 ...........................................................

**b** Explain how the structure of graphite is different from the structure of diamond. Explain why this difference in structures gives graphite the different properties you have listed in your answer to part **a**. You can use a diagram in your answer if you wish.

.................................................................................................................................

.................................................................................................................................

.................................................................................................................................

.................................................................................................................................

## ② Silicon dioxide

Silicon dioxide is a common compound in the crust of the Earth.

**a** Complete the sentences by filling in the blanks and putting a ring around the correct bold words.

Quartz is one of the crystalline forms of ........................................................... .

Silicon dioxide is **hard / soft**, it **conducts / doesn't conduct** electricity, and it has a **high / low** melting point.

**b** Colour this diagram of the structure of quartz. Colour the oxygen atoms *red* and the silicon atoms *grey*. Then complete the labels.

This is an example of a

........................................... structure.

Each oxygen atom forms

............... covalent bonds.

Each silicon atom forms

............... covalent bonds.

○ Si atoms ◯ O atoms

The strong bonds between the atoms

are ........................................... bonds.

There are two oxygen atoms for

every silicon atom, so the

formula is ........................................... .

# Metals from the lithosphere

## 1 Metals and metal ores

Complete the questions on this page with the help of the two tables.

**Table 1**

| Metal | Metal ore | Formula of the mineral in the metal ore |
|---|---|---|
| aluminium | bauxite | $Al_2O_3$ |
| iron | magnetite | $Fe_3O_4$ |
| potassium | sylvite | KCl |
| tin | cassiterite | $SnO_2$ |
| zinc | zincite | $ZnO_2$ |

**Table 2**

| Reactivity | Metal |
|---|---|
| most reactive | K |
| The more reactive a metal is, the more strongly it holds on to oxygen and the more difficult it is to extract the metal. | Al |
| | Zn |
| | Fe |
| | Sn |
| | Cu |
| least reactive | Au |

**a  i** Name an element from Table **2** that can be found free in nature.

.......................................................................................................................................................................

**ii** Why is this element found uncombined with other elements?

.......................................................................................................................................................................

**b** Name three metals in Table **1** that can be extracted from their oxide ores by heating with carbon.

1 ................................... 2 ................................... 3 ...................................

**c  i** Name two metals in Table **1** that cannot be extracted by heating with carbon.

1 ................................................... 2 ...................................................

**ii** Give a reason for your choices.

.......................................................................................................................................................................

**iii** Name the method used to extract these metals. ...................................................................

## ② Calculating percentages of metal in metal ores

**a** Magnetite is a type of iron ore. It contains iron oxide. Iron can be extracted from iron oxide by heating with carbon.

    **i** Add the missing state symbols and balance this equation for the reaction.

$$Fe_3O_4(s) + \quad C(\underline{\quad}) \rightarrow \quad Fe(s) + \quad CO_2(\underline{\quad})$$

- The chemical that is reduced is ........................................ .
- The chemical that is oxidised is ........................................ .

    **ii** Complete this diagram to work out the percentage of iron in iron oxide, $Fe_3O_4$.

| Relative atomic masses: Fe = 56      O = 16 |
|---|

The formula

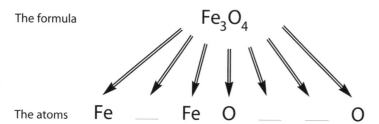

The atoms     Fe  ..........  Fe   O  ..........  ..........  O

The relative  ..........  ..........  ..........  ..........  ..........
atomic masses

The relative formula mass of the iron oxide = ..................................

In this formula there are ..................... atoms of iron, Fe.

The relative mass of ........................... Fe = ........................... .

This means that in ................ kg of $Fe_3O_4$ there are ................ kg of Fe.

So 1 kg of $Fe_3O_4$ contains ........................... kg of Fe.

So 100 kg of $Fe_3O_4$ contains ........................... kg of Fe.

Another way of saying this is that the percentage of Fe in $Fe_3O_4$ = ........................... %

**b** Work out the percentage of copper in bornite by filling in the gaps to show your working.

| Relative atomic masses: Cu = 64     Fe = 56     S = 32 |
|---|

The relative formula mass of bornite, $Cu_5FeS_4$ = ..............................

Mass of copper in the relative formula mass = ..............................

This means that in ..................... kg of $Cu_5FeS_4$ there are ..................... kg of Cu.

So 1 kg of $Cu_5FeS_4$ contains ........................... kg of Cu.

So 100 kg of $Cu_5FeS_4$ contains ........................... kg Cu.

So the percentage of Cu in $Cu_5FeS_4$ ........................................... %

**c** What mass of lead could be extracted from 100 kg of lead oxide ($PbO_2$)?

> Relative atomic masses: $Pb = 207$    $O = 16$

...........................................................................................................................

...........................................................................................................................

**d** What mass of zinc could be extracted from 1 tonne of zinc sulfide (ZnS)?

> Relative atomic masses: $Zn = 65$    $S = 32$

...........................................................................................................................

...........................................................................................................................

## ③ Electrolysis of aluminium oxide

**a** Label the diagram to describe the equipment used to extract aluminium from aluminium oxide. Use these words and phrases.

| | | |
|---|---|---|
| carbon anodes | carbon lining | negative electrode |
| molten aluminium oxide | molten aluminium | tapping hole |

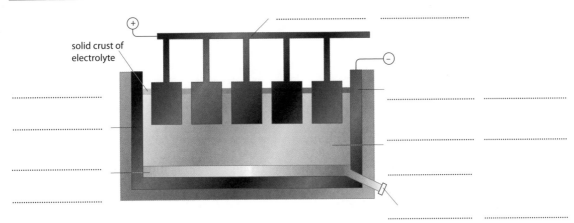

**b** Explain why aluminium oxide conducts electricity when liquid but not when solid.

...........................................................................................................................

...........................................................................................................................

**c** Write the symbols for the two ions in aluminium oxide. ....................................................

**d** Complete this equation to show what happens at the negative electrode during the electrolysis of molten aluminium oxide.

.......................... $+ 3e^- \rightarrow Al$

**e** Complete these equations to show what happens at the positive electrode during the electrolysis of molten aluminium oxide.

.......................... $\rightarrow O + 2e^-$

.......................... $+$ .......................... $\rightarrow O_2$

## ④ Electrolysis of sodium chloride

**a** Complete the labelling of the diagrams to explain what happens during the electrolysis of molten sodium chloride. Choose from these words.

| | | | | | |
|---|---|---|---|---|---|
| melts | conduct | chloride | atoms | ions | metal |
| positive | move | ions | molecules | conductor | electrons |

The ............................................. in solid sodium chloride cannot move around,

so salt does not ............................................. electricity.

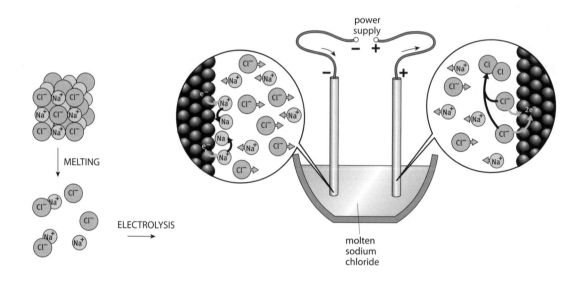

When sodium chloride

............................. the ions can

............................. . So molten sodium

chloride is a ............................. .

Positive sodium ............................. move to the negative electrode.

Here they gain ............................. and turn into uncharged atoms of

sodium ............................. .

Negative .............................

ions move to the ............................. electrode. Here they lose their

extra electron and turn into

chlorine atoms. Chlorine atoms

pair up to make .............................

of chlorine gas.

**b** Write equations for the reactions that occur at the positive and negative electrodes during the electrolysis of molten sodium chloride.

- At the negative electrode: ..................................................................................... .

- At the positive electrode: ................................................. then ................................. .

# H  Structure and bonding in metals

## ① Properties of metals

**a** Read the following paragraph and <u>underline</u> four different properties of metals.

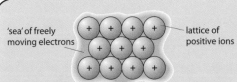

'sea' of freely moving electrons — lattice of positive ions

*A model of metallic bonding*

In a metal, such as copper, the atoms are packed closely together. The atoms are held together by strong metallic bonds, so copper is strong and difficult to melt. Copper is malleable, which means that it can be beaten into a different shape. This is because the atoms can be moved around without the structure losing its strength. When a metal structure is formed, the metal atoms lose their outer electrons and form positive ions. The electrons are no longer held by particular atoms, so they can move freely between the positive ions. This is why metals are good conductors of electricity.

**b** In this table list the four properties you underlined above. In the second column, explain why metals have each property, using ideas about metallic bonding.

| Property | Explanation |
|---|---|
| 1 | |
| 2 | |
| 3 | |
| 4 | |

# The life cycle of metals

## ① Sustainable development

Many people are worried about changes in our environment. They are trying to encourage support for sustainable development.

**a** Draw lines to match each term to its correct meaning.

| sustainable | | Changes in our world.<br><br>For example, changes to how we grow our food, make goods, and organise society. |

| development | | Changes that:<br>• meet everyone's needs<br>• protect the environment<br>• conserve natural resources |

| sustainable development | | Can be done without harming people or the environment. |

**b** Statements A–L are all about metals.

| | | | |
|---|---|---|---|
| **A** Waste iron and steel are put into the reaction vessel together with the raw materials for making new iron and steel. | **B** There is increased use of composite materials, with different materials being bonded together. | **C** Steel cans are on average 10 g lighter than they were 20 years ago. | **D** Make manufacturers responsible for the disposal of the goods they supply. |
| **E** Many metal products can be used for years, so the environmental effects of their production will be less than using less durable materials. | **F** Unsorted household waste is dumped in landfill sites. | **G** Collecting and reprocessing aluminium is the most cost-effective form of recycling. | **H** Only a third of the aluminium thrown away by UK households is recycled. |
| **I** 24 million tonnes of new aluminium are produced worldwide each year. | **J** Batteries contain useful metals. The UK generates about 25 000 tonnes of waste general-purpose batteries each year. Only about 1000 tonnes are recycled. | **K** It is now against the law to mix batteries containing certain metals with general rubbish. | **L** Metals can be recycled indefinitely without losing their properties. |

For each question below, write down the letters (A–L) of the statements that apply.

**i** Which statements are about using metals in ways that are not sustainable?

**ii** Which statements are about making the use of metals more sustainable?

..................................................................................................................................................

**iii** Which statements do you need more information about to decide whether or not they improve sustainability?

..................................................................................................................................................

## 2) The life cycle of a metal

Use these words and phrases to complete the diagram, which shows the life cycle of a metal. Two of the boxes have been filled in for you.

| | | |
|---|---|---|
| metal in use | rubbish to waste tip | |
| processing the ore to produce pure mineral | separating and recycling waste metal | recycling scrap metal |
| making products from the metal | extracting the metal from the mineral | |
| end of useful life | | |

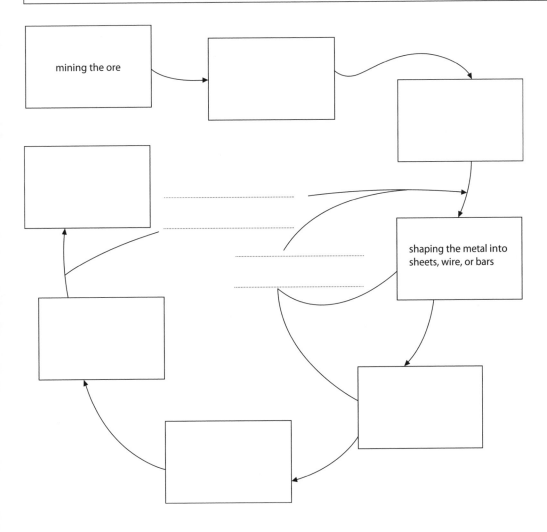

mining the ore

shaping the metal into sheets, wire, or bars

### ① Static

Insulating materials get charged up when they are rubbed. This is due
to stationary charge. It is called static electricity.

**a** A piece of polythene is rubbed with a cloth and gets charged. Another piece
of polythene is rubbed with the cloth.

**i** Complete these sentences. Draw a ring around the correct **bold** words.

The two pieces of polythene will get **the same / different** charge. This means
that they will **attract / repel** each other. The polythene was charged because
of the movement of **electrons / ions,** which have a **positive / negative** charge.

**ii** Look at the picture below. It shows that the polythene became negatively charged.
There is a line that shows the transfer of electrons. Draw an arrow on one end
of this line to show the direction in which the electrons moved.

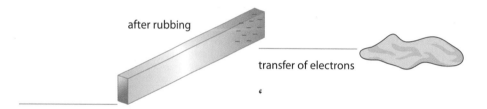

after rubbing

transfer of electrons

**iii** Explain how the neutral cloth has become positively charged. ........................................

.................................................................................................................................

.................................................................................................................................

**iv** There is a force between charges. Draw a line
to match the start of each sentence with
its correct ending.

| Like charges . . . | . . . attract each other. |
| Opposite charges . . . | . . . repel each other. |

**b** Polythene, perspex, and nylon all get charged when they are rubbed. Polythene
gets a negative charge. Perspex gets a positive charge.

**i** Fill in the blanks to complete these sentences.

A piece of perspex will ........................................... another piece of perspex.

A piece of perspex will ........................................... a piece of polythene.

**ii** Polythene repels charged nylon. What charge must nylon have? ........................................

**iii** What effect will perspex have on nylon? ........................................

## ② Charge and discharge

The dome of a Van de Graaff generator gets charged up when it is switched on.
Yasmin is holding the dome and her hair stands on end.

**a** Explain why her hair stands on end.

.......................................................................................................................................................................

**b** The teacher discharges the dome and Yasmin's hair falls back down. Describe what happens to Yasmin's charge and why her hair falls back down.

Use the words charge and neutral.

.......................................................................................................................................................................

.......................................................................................................................................................................

.......................................................................................................................................................................

## Electric currents in circuits

### ① Rope model of an electric circuit

The picture shows a model of an electric circuit. Bethan is the 'battery'. She pulls the loop of rope around in the direction shown by the arrows. The other pupils let it pass through their hands.

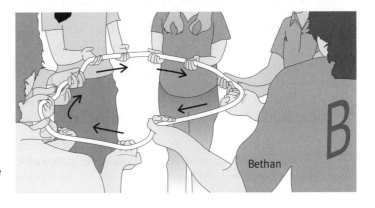

Bethan

Join the boxes on the next page to show how the model helps to explain an electrical circuit. The first one has been done for you.

| Rope circuit | Electrical circuit |
|---|---|
| When Bethan first pulls the rope, it starts moving through everyone's hands at the same time. | Stored energy is transferred out of the battery. |
| Bethan gets tired after pulling the rope around. | Putting in an insulator stops the flow of charge. |
| The others feel their hands getting hot. | The current is not used up. It is the same everywhere. |
| If any one of the others grips the rope firmly, the rope stops moving. | Charge moves throughout the circuit as soon as it is connected up. |
| At any time, the amount of rope leaving each child's hand is the same as the amount going in. | The battery does work on all other components in the circuit. |

## ② Electric current

Physicists think that an electric current is the flow of charge – the same charge that causes static electricity. The statements below explain one piece of evidence for this belief but they are out of sequence.

**i** Use arrows to join up the statements in the correct order. The first one has been done for you.

| This implies that the electric current, | When the sparks jumps across the gap, |
|---|---|

| These are the same charges that jump across the gap to make a spark. | is a movement of charges. |
|---|---|

| which makes the lamp light up, | the lamp lights up. |
|---|---|

**ii** Write 'D' by a box that reports data. Write 'E' by the three boxes that, together, are an explanation.

## ③ Electrons

All matter is made of atoms, which contain protons (positive) and electrons (negative). In metals, the electrons are free to move throughout the metal. That is why metals are electrical conductors.

The flow of charge around a circuit is called electric current. Traditionally, physicists and engineers use 'conventional current', showing the direction that positive charges would be flowing. The electrons actually flow in the other direction.

**a** Look at the descriptions below. Draw lines to match current to its description in the middle and its direction on the right.

| 'conventional current' | flow of negative electrons | from – terminal to + terminal |
|---|---|---|

| actual movement | imaginary flow of positive charges | from + terminal to – terminal |
|---|---|---|

**b** The diagram shows atoms in a wire.

- Label the end of the wire joined to the positive (+) battery connection.

- Draw an arrow next to the wire to show the direction of 'conventional current'.

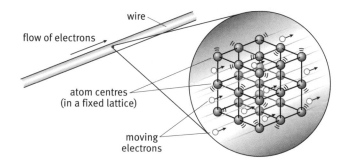

## ④ Computer model of an electric circuit

**a** This diagram shows a computer model of an electric circuit.

Complete the sentences.

The electric current is being measured by an ........................................................ .

The meter reading is 10 milliamps (mA).

This is the same as ................................................ amps (A).

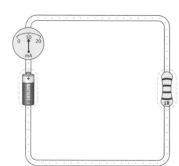

**b** This diagram shows the same circuit, but it now has two extra meters.

Put arrows in each meter to show the readings, then complete the sentences.

The charges in an electric circuit never get used up.

Current is the *flow* of charges, so the current is

.................................................................................................

## Branching circuits

### ① **Current in parallel circuits**

When several components are connected in parallel directly to a battery

**X** • the current through each component is the same as if it were the only component present

**Y** • the total current from (and back to) the battery is the sum of the currents through each of the parallel components

Circuit 1 on the right shows a single resistor connected to a battery.

The current through the resistor is 6 mA.

**a** What is the current coming from the battery?

.................................................................................................

**b** Put arrows in the other two ammeters to show their readings.
In circuit 2, another resistor, resistor B, has been added in parallel.

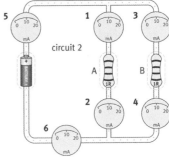

**c** What will the current be in *resistor A* in circuit 2?

.................................................................................................

**d** Which of the statements at the top of the page tells you this?
Draw a ring around the correct statement below:

**statement X**     **statement Y**

**e** Draw arrows in ammeters 1 and 2 to show their readings.

**f** Resistor B is exactly the same as resistor A. What will the current be in resistor B?

.................................................................................................

**g** Draw arrows in ammeters 3 and 4 to show their readings.

**h** What will be the current through the battery?

.......................................................... mA

**i** Which of the statements at the top of the page tells you this?
Draw a ring around the correct statement below:

**statement X**     **statement Y**

**j** Draw an arrow in each of the ammeters 5 and 6 to show their readings.

**k** Now imagine you add a third similar resistor in parallel. Put a tick ✓ next to the true statement below.

- The current through the third resistor will be the same as the currents through resistors A and B and the current coming from the battery will increase. ☐

- The current coming from the battery will stay the same and the current in resistors A and B will drop slightly. ☐

## D    Controlling the current

### ① Resistance

**a** Look at the statements on the left below. Some of them apply to electric current and some apply to electrical resistance.

Draw a line to link each statement to the correct box on the right. The first one has been done for you.

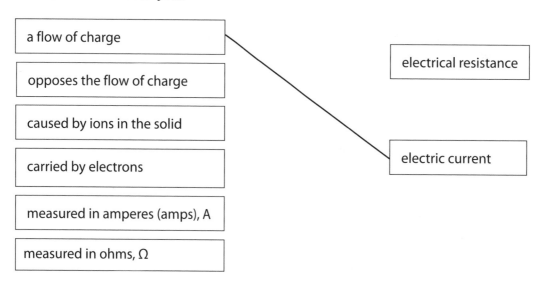

| a flow of charge |
| opposes the flow of charge |
| caused by ions in the solid |
| carried by electrons |
| measured in amperes (amps), A |
| measured in ohms, Ω |

electrical resistance

electric current

**b** Complete the sentences below. Draw a ring around the correct **bold** words.

- The resistance of a circuit determines the **size** / **direction** of the current.

- A big resistance makes it **more** / **less** difficult for the charge to flow and leads to a **big** / **small** current.

- It is easier for charge to flow through a **smaller** / **bigger** resistance so the current through it is **smaller** / **bigger**.

- Insulators have a very **small** / **big** resistance and so the current through them is practically zero.

### ② The cause of resistance

The electrical resistance of metals can change.

**a** Complete this description of the *correlation* between the electric current flowing through a metal and its temperature.

As the electric current ........................................................................................................

........................................................................................................................................

**b** Scientists only say that a factor *causes* an outcome if they can suggest a *mechanism* to explain it.

**i** Fill in the blanks to complete this statement.

In part **a** the *factor* is ........................................... and the *outcome* is ........................................... .

**ii** Explain the *mechanism* involved.

Use the words from the box in your explanation.

| lattice    ions    electrons |
| --- |

...........................................................................................................................

...........................................................................................................................

...........................................................................................................................

## ③ Ohm's law

Use these words to complete the sentences below. Words can be used once, more than once, or not at all.

| battery    bigger    double    proportional    size    smaller    voltage |
| --- |

A ........................................... pushes charge around an electric circuit.

A battery's ........................................... is a measure of its push. The bigger a battery's

voltage, the ........................................... its push on the charge. In turn, this will lead

to a ........................................... current. The current is ...........................................

to the voltage. This means that doubling the voltage will ........................................... the current.

## ④ Measuring resistance

Look at the circuit on the right. It has two 1.5 V batteries giving a total voltage of 3 V. The current in the circuit is 24 mA.

Now look at the circuits below. The resistance of each of the four circuits is the same, but the number of 1.5 V batteries changes.

**a** In each circuit, write the voltage across the resistor next to the voltmeter.

**b** In each circuit, write the current next to the ammeter.

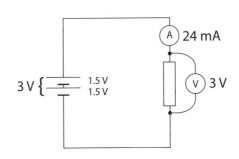

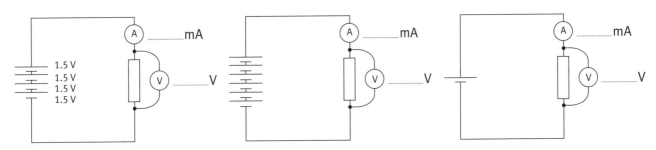

## ⑤ Graphs of current against voltage

The graph on the right shows how the current varies in a fixed resistor as the voltage is increased.

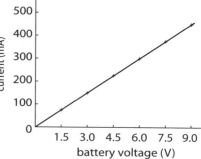

**a** Complete this sentence by filling in the missing word.

The current is ................................................ to the voltage.

**b** The test is done on a resistor with twice the resistance.
What difference will this make to the currents at each voltage?

................................................................................................

**c** Draw in the line you would get for current against voltage with twice the resistance.

**d** The equation below is used to calculate resistance. Complete the equation by putting in the units.

Resistance (................ , Ω) = $\dfrac{\text{voltage (................, V)}}{\text{current (................, A)}}$

**e** When the results for the graph above were collected there was another pair of values: voltage = 10.5 V, current = 420 mA.

**i** Tick ✓ the best reason for not plotting this value.

☐ Six points are enough to show the pattern.

☐ The point does not fit on the axes.

☐ The point does not fit the pattern of the other results.

☐ There was a good reason for deciding that this point was an error in the results.

**ii** These suggestions were made when the graph was being plotted.

• Who is saying what a scientist should do? ................................................

**iii** When the result was checked the voltage = 10.5 V and current = 400 mA.
The circuit had been left switched on and the wires and the resistor were warm.

• Give a reason for deciding the point was an outlier and not plotting it.

................................................................................................

## ⑥ Variable resistors

**a** Some electrical components have a resistance that changes. Look at the three components below left.

   **i** Draw a line from each component to match it to the description of how to change its resistance.

   **ii** Draw a line to join each of the descriptions to a possible use on the right.

    The first one has been done for you.

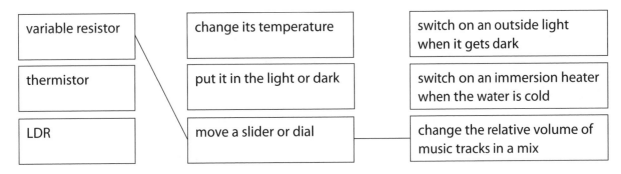

| variable resistor | change its temperature | switch on an outside light when it gets dark |
| thermistor | put it in the light or dark | switch on an immersion heater when the water is cold |
| LDR | move a slider or dial | change the relative volume of music tracks in a mix |

**b** What do the initials LDR stand for? L ........................ D ........................ R ........................

**c** When does the resistance of an LDR increase? ........................

**d** When does the resistance of the most common type of thermistor increase? ........................

## ⑦ Combining resistors

**a** Combining resistors in series and parallel will make a new resistance.

   **i** Complete the statements below (left). Draw a ring around the correct **bold** words.

   **ii** Draw a line to link each statement to the correct explanation on the right.

| Two resistors in series have a **bigger / smaller** resistance than either one on its own. | Because there are more paths that the moving charges can take. |
| Two resistors in parallel have a **bigger / smaller** resistance than either one on its own. | Because the moving charges have to pass through one then the other. |

**b** Look at the networks of resistors below. Each resistor has the same value. Put them in order of increasing resistance. Number them from 1 to 5: the smallest has the number 1 and the largest is 5. The first one has been done for you.

## ⑧ Circuit symbols

Label the circuit symbols with their names.

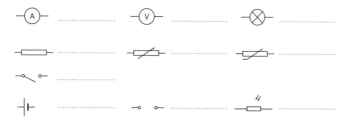

---

## ① Potential energy

The voltage of a battery is the 'push' the battery gives the charge. But it is also the work the battery does in pushing each unit of charge around the circuit.

Inside the battery, the chemical reactions give each charge some *potential energy*.

The water model can help to show this. The pump pushes on the water, rising it up and increasing its potential energy. It loses this potential energy as it flows round the circuit.

The battery does work on the electric charges.

The electric charges do work on the atoms in the resistor.

The pump does work in lifting the water.

The water has gained potential energy.

PUMP

The water loses the energy as it falls back down.

**a** Join the boxes below to show how the model helps to explain the energy changes in an electrical circuit. The first one has been done for you.

| | |
|---|---|
| The battery pushes on the charge, raising its potential energy. | The water loses potential energy as it falls into the tray. |
| The resistor heats up as the moving charge does work on its atoms. | The pump pushes on the water, raising it up to where it has more potential energy. |
| The charge loses potential energy as it does work in the resistor. | The water heats up slightly when it splashes into the tray. |
| The potential energy lost by the charge in the resistor is the same as the potential energy it gained in the battery. | The water loses the same potential energy when it falls into the tray as it gained in the pump. |

**b** As charge flows around a circuit, its potential energy changes. The charge gains potential energy in a battery and loses potential energy in resistors (or other components).

Fill in the blanks to complete the sentences below.

**i** The p ............................................... d ............................................... between two points

in a circuit is measured in ............................................... .

**ii** Voltage is another word for .............................................................................................................................. .

**iii** The bigger the potential difference between two points in a circuit, the

more .............................................................. is transferred between these points.

## 2) Combining potential differences

**a** Look at the circuits on the right. The batteries and lamps are identical.
The only difference is that circuit B has two batteries in parallel.

Look at the comparisons in the table below.
Put a tick ✓ in the column that correctly describes
how they compare in the two circuits.

circuit A

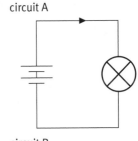

|  | the same | bigger in A | bigger in B |
|---|---|---|---|
| the potential difference across the bulb |  |  |  |
| the current through the bulb |  |  |  |
| the current through a single battery |  |  |  |
| the time for the batteries to go flat |  |  |  |

circuit B

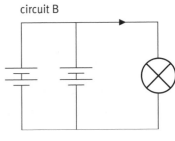

**b** Potential differences add up around a circuit.

Look at the circuit on the right and the
statements below. Draw a line to match each
of the statements on the left to its explanation
on the right. The first one has been done for you.

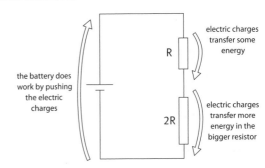

| The bigger the potential difference across the battery, the bigger its push. | Energy is conserved in the circuit: there is no net gain or loss of energy. |
|---|---|
| There is a drop in potential across each of the resistors. | The harder the battery pushes the charge, the more work it does. |
| There is a bigger potential difference across the bigger resistor. | The charge does work as it moves through a resistor. |
| The sum of potential differences across two resistors equals the potential difference across the battery. | The charge does more work as it goes through a bigger resistor. |

## 3) Current and p.d. in series circuits

Complete the sentences below. Draw a ring around the correct **bold** words.

When two resistors are in series:

**a** the current through the bigger resistor will be **bigger than** / **the same as** /
**smaller than** the current in the smaller resistor.

**b** The potential difference across the bigger resistor will be **bigger than** / **the same as** / **smaller than** the potential difference across the smaller resistor.

## ④ The potential divider

Look at the circuits below. They are called *potential divider* circuits.

**a** For each circuit, put a tick ✓ next to the resistor that has the bigger potential difference across it.

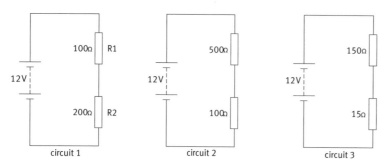

circuit 1      circuit 2      circuit 3

**b** In circuit 1, the total resistance is 300 Ω.

Calculate the current in the circuit. Use the equation current = voltage/resistance.

current $I$ = $\dfrac{V}{R}$ = $\dfrac{\text{..................................} \ V}{\text{..................................} \ \Omega}$ = .................................. A

**c** Calculate the potential difference across each of the resistors in circuit 1.

Use the equation voltage = current × resistance.

p.d. across R1 = .................................. A × .................................. Ω = .................................. V

p.d. across R2 = .................................. A × .................................. Ω = .................................. V

**d** Does this agree with the tick you put on circuit 1 in part **a**? ..................................

**e** Explain why the sum of the potential differences in part **c** must be 12 V.

..................................................................................................................................................

## ⑤ Electronic control

To keep the temperature constant in a greenhouse, a gardener uses an electronic circuit to switch the heater on and off automatically.

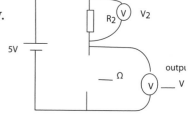

**a i** The circuit contains a thermistor. Complete the sentence to explain the choice of sensor.

When the temperature increases the ..................................

of the thermistor .................................. .

**ii** Add the symbol for a thermistor to this circuit diagram of the control circuit.

**b** The circuit must produce a voltage to switch on the heater. The heater will switch on if the voltage is above 3 V and off if it is below 3 V.

The gardener wants this to happen at a temperature of 12 °C. At 12 °C the resistance of the thermistor is 30 kΩ.

  **i** Complete this statement.

  As the temperature decreases the potential difference across the

  thermistor, $V_1$ will ........................................................

  **ii** Use the information above to fill in the values on the circuit diagram for:

  - the resistance of the thermistor when the heater must switch on
  - the potential difference across the thermistor when the heater must switch on

**c** From the information on the circuit diagram:

  **i** Work out, and write on the circuit diagram, the value of the potential
  difference $V_2$ when the heater must switch on.

  **ii** what resistance must the resistor $R_2$ have? (Ring) the correct value.

| | | | |
|---|---|---|---|
| 20 kΩ | 30 kΩ | 45 kΩ | 50 kΩ |

## ⑥ Current in parallel branches

**a** Complete the statements below. Draw a (ring) around the correct **bold** words.

  When two components are in parallel:

  - the potential difference (voltage) across each one is **the same** / **different**
  - the potential difference across each one is **equal to** / **smaller than** the
    voltage of the battery

**b** A hairdryer has a heater and a fan motor in parallel. There are two
  switches. When switch **A** is closed, the fan comes on. When switch
  **B** is closed, the heater comes on as well. The heater will not come
  on unless switch **A** is already closed.

  **i** Complete the circuit diagram below right by labelling the
  two switches **A** and **B**.

  **ii** Suggest why it is important that the hairdryer cannot
  be turned on with the heater on and the fan off.

  ..............................................................................

  ..............................................................................

  ..............................................................................

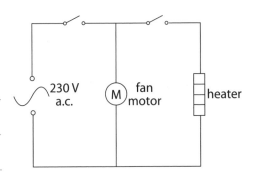

  **iii** The motor has a resistance of 460 Ω and the heater has a resistance of
  46 Ω. The mains supply is 230 V. Rearrange the equation $R = \frac{V}{I}$ to calculate:

  - the current in the 46 Ω heater

  current $I$ = $\dfrac{\text{........................}}{\text{........................}}$ = $\dfrac{\text{........................ V}}{\text{........................ Ω}}$ = ........................ A

- the current in the 460 Ω motor

current $I$ = _____ = $\dfrac{\text{_____ V}}{\text{_____ Ω}}$ = _____ A

- the total current from the 230 V supply = _____ A

**c** Complete these sentences. Draw a ring around the correct **bold** words.

When two components are in parallel, the current in each one is **the same as** / **smaller than** the current for that component on its own. The total **current** / **voltage** taken from the power supply will be **more** / **less** than it was with just one component on its own. This extra current will have a cost: the time for the batteries to go flat will **increase** / **stay the same** / **decrease**.

**d** Look at the list of words below. Draw a <u>line</u> under those that go *through* a component and a ring around those that go *across* a component.

| current | flow | potential difference | voltage |
|---------|------|----------------------|---------|

---

## F    Electrical power

### ① Power, voltage, and current

**a** In electric circuits, the equation for power is:

power           = potential difference (voltage)    ×      current

( _____ )         ( _____ )         ( _____ )

   **i** Complete the equation above by putting in the units.

A 60 W bulb is connected to the mains. The mains voltage is 230 V.

   **ii** Calculate the current through the bulb when it is working normally. (You will need to rearrange the equation to find the current.)

current = ⬚ = _____ A

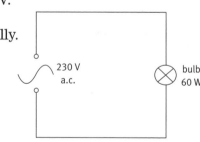

**b** Power measures the rate at which energy is transferred from the supply. Complete the sentences below.

A 100 W light bulb is _____ than a 60 W bulb. It transfers

energy _____ quickly. They are both connected to the same

mains voltage of _____ . This means that _____

charge has to flow through the 100 W bulb each second. So the current in the 100 W

bulb is _____ .

## Magnets and motors

### ①  Magnetic forces

**a**  Draw a ring around the correct **bold** words in this statement.

When an electric current flows through a wire there is a magnetic **field** / **fluid** in the region **around** / **above** the wire.

**b**  Tick ✓ all the correct statements.

When a current-carrying wire is placed in a magnetic field there is a force on the wire. The force is always:

☐  along the wire                     ☐  downwards

☐  at right angles to the magnetic field      ☐  towards the magnetic north pole

☐  at right angles to the wire         ☐  upwards

### ②  An electric motor

The diagram shows a simple electric motor.

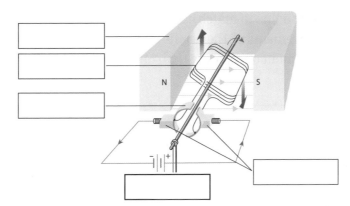

**a**  Use the words from the list to complete the labels on the diagram.

| battery    brushes    coil    commutator (split ring)    magnet |

**b**  Explain why the horizontal coil turns when a current passes through it.

.................................................................................................................

.................................................................................................................

**c**  The direction of the current is reversed twice every turn.

**i** What position is the coil in when the current reverses? ...........................

**ii** Explain how the current is reversed.

.................................................................................................................

.................................................................................................................

**d** Draw one line to join each change to its effect on the rotation of the motor.

**Change**

| |
|---|
| swap the position of the N and S pole of the magnet |

| |
|---|
| swap the position of the + and − terminals of the battery |

| |
|---|
| swap the N and S pole of the magnet AND the + and − terminals of the battery |

**Effect on rotation**

| |
|---|
| reverses |

| |
|---|
| no change |

---

**H** **Generating electricity**

① **Generator effect**

**a** A moving magnet induces a voltage in a coil of wire. The picture shows the north pole of a magnet moving into a coil. The needle on the ammeter flicks to the right.

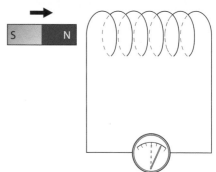

Look at the boxes below. Draw a line to match each of the actions on the left with one of the movements of the needle on the right. You can use each needle movement once, more than once, or not at all.

This ammeter reads zero when its needle is in the middle.

| |
|---|
| pull the north pole out of the coil |

| |
|---|
| hold the magnet stationary in the coil |

| |
|---|
| push the south pole into the coil |

| |
|---|
| pull the south pole out of the coil |

| |
|---|
| flicks to the left |

| |
|---|
| no movement |

| |
|---|
| flicks to the right |

**b** The magnet can be put on a spindle and rotated near the coil. This will induce an alternating current in the coil.

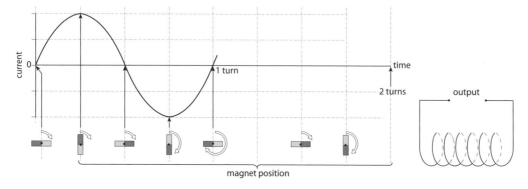

magnet position

This diagram shows how the induced current varies as the magnet rotates through different positions. It shows one cycle of the a.c.

   **i** Complete the diagram, showing how the current varies through two full rotations of the coil.

   **ii** Draw in the missing magnet positions.

**iii** Write down four ways in which you could induce a bigger voltage in the coil.

....................................................................................................    ....................................................................................................

....................................................................................................    ....................................................................................................

## Distributing electricity

### (1) Transformers

Look at the picture of a simple transformer.

**a** Complete the diagram using these labels.

| iron core | primary coil |
|-----------|--------------|
| induced alternating current | |
| secondary coil | a.c. supply |

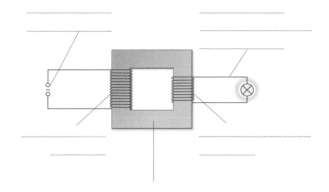

**b** Complete these sentences. Draw a ring around the correct **bold** words.

A transformer works because the current in the **primary / secondary** coil
produces a **magnetic / electric** field that passes through the secondary coil.
This field is changing and therefore **induces / reduces** a voltage.

The transformer above has **more / fewer** turns on the secondary coil. This means it is a **step-up /
step-down** transformer. The output voltage will be **more / less** than the input voltage.

**c** This is the transformer equation: $\dfrac{V_p}{V_s} = \dfrac{N_p}{N_s}$

The input voltage is 230 V, and the number of turns are 460 (primary) and
24 (secondary). What is the output voltage from the transformer?

$$V_s = \text{................................................} \ V$$

**d** The National Grid distributes mains electricity from power stations to people's homes.

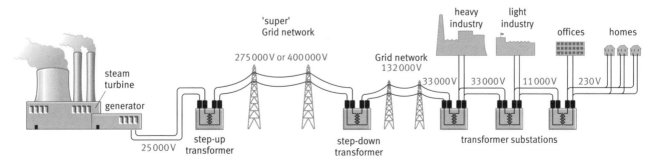

Look at the phrases below. Some of them apply to a.c. and some to d.c.
Put a <u>straight line</u> under the phrases that refer to d.c. and a <u>wavy line</u> under
the phrases that refer to a.c.

It is easier to generate.     Its voltage has a constant value.     It can be distributed more efficiently.

It is produced by           It comes from the mains supply.       It won't pass through a transformer.
batteries.

## What is behaviour?

### ① Simple reflexes

Animals respond to stimuli in order to keep themselves in favourable conditions.

**a** A stimulus is a change in the environment of an organism. Look at the examples of behaviour below. For each example write down:

  **i** the stimulus

  **ii** the response.

> Woodlice prefer dark places; they move away from light.
>
> Bacteria living in the gut move towards the highest concentration of food.
>
> An earthworm rapidly withdraws into its burrow if pecked.
>
> A resting housefly takes off as soon as it sees any fast movement nearby.
>
> If an octopus sees a predator, it releases a cloud of 'ink' and moves away quickly.

**b** Most of the behaviour of simple animals is made up of reflex actions. This means they cannot adapt their behaviour, or learn from experience. Explain the disadvantages of this.

.................................................................................................................................

.................................................................................................................................

.................................................................................................................................

### ② a  Put a ring around each correct word to describe a simple reflex.

| Simple reflexes produce | immediate | and | planned | responses to a | stimulus. |
|---|---|---|---|---|---|
| | fast | | voluntary | | reflex. |
| | slow | | involuntary | | receptor. |

**b** Explain why you have chosen each of the three words.

Word 1 ....................................................................................................................

.................................................................................................................................

Word 2 ....................................................................................................................

.................................................................................................................................

Word 3 ....................................................................................................................

.................................................................................................................................

## Simple reflexes in humans

### ① Human reflexes

Simple reflexes produce rapid involuntary responses.

**a** Simple reflexes are responses that you do not think about or learn. Look at the list below. Highlight or <u>underline</u> the examples of simple reflexes.

- Your pupils get smaller in bright light.
- You answer a question.
- Your eyes water on a windy day.
- A newborn baby grasps at anything put in her hand.
- A goalkeeper saves a long-range shot at goal.
- You breathe faster when you run.

**b** Humans and other mammals have very complex behaviour, but simple reflexes are also important for their survival. Complete the table to describe reflex actions in adults and newborn babies.

| Adult reflex | Stimulus | Response |
|---|---|---|
|  |  |  |
|  |  |  |
|  |  |  |

| Newborn reflex | Stimulus | Response |
|---|---|---|
|  |  |  |
|  |  |  |
|  |  |  |
|  |  |  |

**c** Many newborn reflexes are present for only a short time after birth. Explain why they increase a young baby's chances of survival. Use these words in your explanation.

| behaviour | experience | learn |
|---|---|---|

..................................................................................................................................

..................................................................................................................................

..................................................................................................................................

② Simple reflexes involve effectors, a processing centre, and receptors.

a  Explain how each of these is involved when you are surprised as someone taps you on the shoulder from behind.

......................................................................................................................

......................................................................................................................

......................................................................................................................

b  Receptors can be **single cells** or **complex organs**.

Write down an example of each and describe the job that they do.

......................................................................................................................

......................................................................................................................

......................................................................................................................

c  Responses are coordinated by both the nervous system and the hormone system.

Complete the table to describe the differences between the two systems.

| Differences | |
| --- | --- |
| Nervous system | Hormone system |
| | |
| | |
| | |
| | |

d  The evolution and development of a nervous system and a hormone system depended upon which of the following?

Put a tick (✓) in the box next to the best answer.

| | |
| --- | --- |
| asexual reproduction | |
| a constant environment | |
| a poor supply of food | |
| the evolution of multicellular organisms | |
| lots of different species | |

## Your nervous system

### ① The peripheral and central nervous systems

Responses are coordinated by the central nervous system (CNS). Sensory and motor neurons carry the signals.

**a** In mammals the nervous system is made up of a **central nervous system** (the **brain** and **spinal cord**). It is connected to the body via the **peripheral nervous system**.

Label or colour the diagram to show the parts of the human nervous system printed in **bold**.

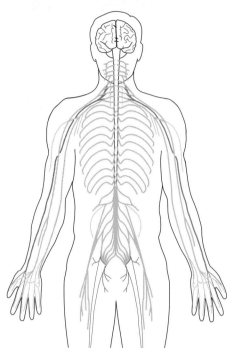

**b** Neurons are cells in the nervous system that carry nerve impulses. Use these words to label the diagram of a neuron.

| axon | cell membrane | cytoplasm |
| fatty sheath | nucleus | |

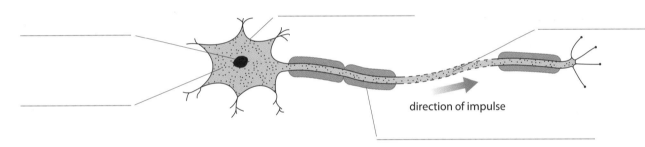

direction of impulse

**c** Neurons transmit electrical impulses when stimulated. Use these words to complete the sentences.

| axon | electrical | insulates | speed |

When the neuron is stimulated, an ........................... impulse travels

along the ........................... to the branched ending. Here it connects

with another neuron or an effector. Some axons are surrounded by a fatty sheath,

which ........................... them from neighbouring cells and

increases the ........................... of the nerve impulse.

**d** In the situation where your pupils contract in bright light:

   **i** what is the stimulus? ...........................

   **ii** where are the receptors? ...........................

**e** **Sensory** neurons carry impulses from **receptors** to the CNS. **Motor** neurons carry impulses from the CNS to **effectors**.

Label the diagram using the **bold** words.

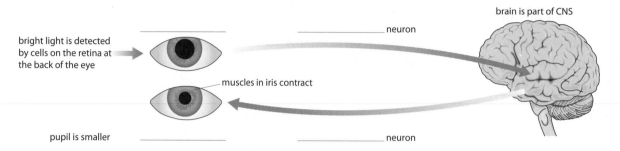

bright light is detected by cells on the retina at the back of the eye

.................................... neuron

brain is part of CNS

muscles in iris contract

pupil is smaller

.................................... neuron

**f** The reactions in parts **d** and **e** are reflex actions. Reflexes are involuntary actions. Many are coordinated in the spinal cord without involving the brain. The route of the message carried along neurons is called a reflex arc. The diagram below shows the reflex arc involved when you step on something sharp. It shows the neurons involved – including the relay neuron in the spinal cord. Add the labels in the box to the diagram.

| effector | motor neuron | receptor | relay neuron | sensory neurone |
| --- | --- | --- | --- | --- |

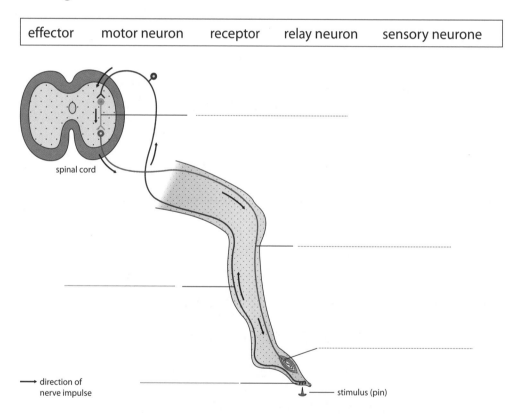

spinal cord

→ direction of nerve impulse

stimulus (pin)

**g** Place the following words in the correct order starting from the stimulus in a reflex arc.

| effector | motor neuron | receptor | relay neuron | sensory neurone |
| --- | --- | --- | --- | --- |

...............................................................................................................................

**h** Which of the following are found inside the spinal cord?

Put a tick (✓) in the boxes next to the correct answers.

| | |
|---|---|
| effector | |
| motor neuron | |
| receptor | |
| relay neuron | ✓ |
| sensory neuron | |

**i** Put a ring around the correct words that best describe the advantage of using a reflex action.

The arrangement of neurons in a

| changing |
| random |
| fixed |

pathway allows reflex responses to be

| automatic. |
| conscious. |
| considered. |

This makes them very rapid since

| no |
| some |
| all |

processing of

| hormones |
| information |
| sensors |

is required.

## Synapses

### ① How synapses work

Chemicals released into the synapses transmit nerve impulses from one neuron to the next.

**a** Synapses are tiny gaps between neurons. Electrical impulses cannot jump across synapses. Chemicals carry impulses between neurons.

Use the diagrams of an impulse crossing a synapse to put the events below in order. Number the boxes 1 to 5. The first one has been done for you.

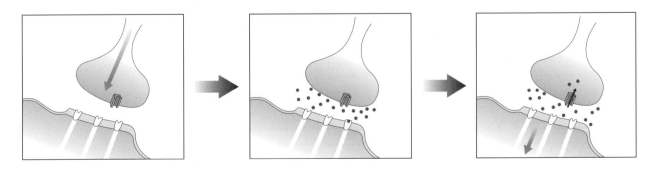

☐ The chemical is absorbed back into the sensory neuron, to be used another time.

☐ An impulse is stimulated in the motor neuron.

[1] A nerve impulse travels along a sensory neuron until it reaches a synapse.

☐ The molecules diffuse across the synapse. They bind to receptor molecules on the membrane of the motor neuron.

☐ The end of the sensory neuron releases a chemical into the synapse.

**b** The receptor molecules only bind to certain chemicals. Complete the diagrams.

   **i** Add receptors of the correct shape to synapse **A**.

   **ii** Add chemicals carrying the impulse to synapse **B**.

   **iii** Add arrows to show the direction of the nerve impulse across synapse **B**.

A                B

synapse A            synapse B

## ② Drugs and toxins

Some drugs and toxins affect the way impulses cross synapses.

**a** Use colours or lines to match up these words with their descriptions. One has been done for you.

| | |
|---|---|
| beta blockers | a poison that causes dangerous effects in the body |
| curare | a medicine or other substance that causes effects in the body |
| drug | drug that stops the heartbeat from speeding up |
| Ecstasy | an antidepressant drug |
| Prozac | a substance that increases nervous activity |
| painkiller | a substance that reduces the sensation of pain |
| stimulant | a poison that blocks transmission of nerve impulses (and stops breathing) |
| toxin | a drug that has mood-enhancing effects |

**b** Highlight or colour the drug that would stop the transmission of an impulse in the synapse shown in the diagram.

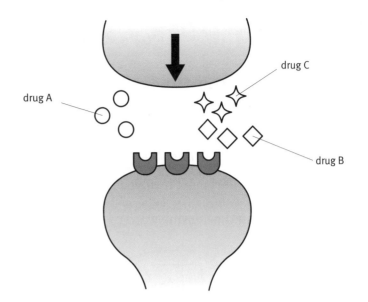

drug C

drug A

drug B

**c** Explain how this drug would stop the nerve impulse from crossing the synapse.

.......................................................................................

.......................................................................................

.......................................................................................

**d** Some drugs work by blocking the reuptake of the synapse chemical. Serotonin is released at one type of synapse in the brain. It triggers nerve impulses causing feelings of pleasure. The drug Ecstasy (or MDMA) blocks the reuptake of serotonin.

Look at the diagram and answer the questions.

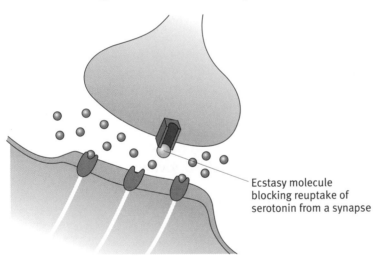

Ecstasy molecule blocking reuptake of serotonin from a synapse

**i** What effect does the presence of the Ecstasy molecule have on the amount of serotonin in the synapse?

.......................................................................................

**ii** What effect does this have on the receptor molecules detecting serotonin?

.......................................................................................

**iii** What effect does this have on the activity of this neuron?

.......................................................................................

**e** Complete the sentence.

The mood-enhancing effects of Ecstasy are due to the ............................................ in serotin concentration at synapses in the brain.

## ① The human brain

The cerebral cortex is a part of the brain. It is most concerned with intelligence, memory, language, and consciousness.

**a** The diagram shows the inside of a human brain.

   **i** Colour the part most involved in intelligence, memory, language, and consciousness.

   **ii** Then fill in the missing words.

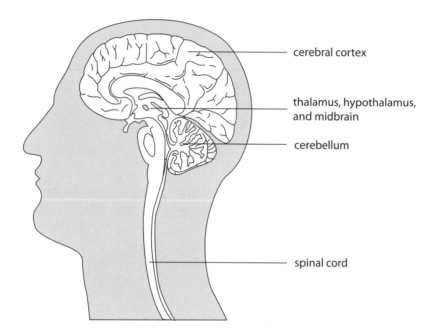

cerebral cortex

thalamus, hypothalamus, and midbrain

cerebellum

spinal cord

The ................................ ................................ is highly folded, giving

it a ................................ surface area. Different areas are responsible

for different ................................. It is much bigger in ................................

than in other mammals when compared to body size.

**b** Studies of patients whose brains have been partly destroyed by injury or disease tell us about the functions of different areas of the cortex.

| Part of cerebral cortex | Function |
|---|---|
| **A** speech centre | talking |
| **B** sensory cortex | receiving information from receptors |
| **C** motor cortex | voluntary movement |
| **D** visual cortex | detecting visual stimuli |
| **E** language | understanding language |

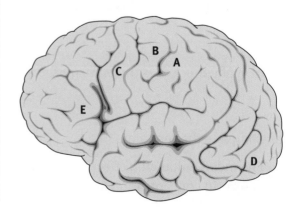

Use the information in the table to decide which area of the brain has been damaged in stroke patients with the symptoms shown in the table.

| Symptom | Damaged brain region (A to E) |
|---|---|
| speech is slurred | |
| has trouble controlling hand movements | |
| legs feel numb | |

c   Some toxins and drugs can affect synapses within the brain.

One of these drugs is Ecstasy.

Name two others.

1   ....................................................................

2   ....................................................................

d   Ecstasy can affect synapses within the brain.

Describe and explain the effect Ecstasy has on the synapses and the subsequent effects on the nervous system.

.............................................................................................................

.............................................................................................................

.............................................................................................................

.............................................................................................................

e   Scientists have studied the brain using patients with brain damage, electrical stimulation, and MRI scans.

Explain how each of these three techniques can improves scientists' understanding of the human brain.

Patients with brain damage ...........................................................

.............................................................................................................

.............................................................................................................

Electrical stimulation ......................................................................

.............................................................................................................

.............................................................................................................

MRI scans ........................................................................................................................................

........................................................................................................................................

........................................................................................................................................

**f** One or more of the techniques mentioned in part e involve ethical issues.

   **i** Explain what these ethical issues are.

   ........................................................................................................................

   ........................................................................................................................

   ........................................................................................................................

  **ii** Summarise different views that might be held about this issue.

   ........................................................................................................................

   ........................................................................................................................

 **iii** Use this issue to explain why the right decision may be the one that leads to the best outcome for the majority of people involved.

   ........................................................................................................................

   ........................................................................................................................

   ........................................................................................................................

  **iv** Use the example to explain why some people may consider that the action is wrong whatever the consequences.

   ........................................................................................................................

   ........................................................................................................................

## Learnt behaviour

### ① Conditioned reflexes

Conditioned reflexes can be learned. The final response has no direct connection to the stimulus.

**a** Animals can learn to link a new stimulus with a reflex action. This is a conditioned reflex response. Pavlov showed this in experiments with dogs. Read what he did and answer the questions.

Pavlov's dog salivated when presented with food.

Pavlov rang a bell while his dog was eating its food.

After a while the dog salivated when it heard the bell, even if no food was around.

  **i** What was the primary stimulus (that originally caused salivation)? _____

  **ii** What was the reflex response? _____

  **iii** What was the secondary stimulus (that Pavlov added)? _____

  **iv** What was the conditioned reflex response? _____

**b** Complete the sentences.

Salivation is a _____ response connected to food. Salivation

and hearing a bell have no direct connection. In a _____ reflex,

the response has no direct connection to the stimulus.

**c** Conditioned reflexes can increase chances of survival. Read the text in the box and answer the questions.

> Many birds feed on caterpillars. Some brightly coloured caterpillars taste nasty to the birds, and could be poisonous.
> Young birds try eating the caterpillars and learn that some taste nasty. In future they avoid all brightly coloured caterpillars.

  **i** What was the primary stimulus? _____

  **ii** What was the secondary stimulus? _____

  **iii** How does this conditioned reflex help the birds to survive? _____

**iv** Hoverflies with markings that make them look like wasps have an increased chance of survival. Explain why.

.................................................................................................

.................................................................................................

.................................................................................................

## (2) Conscious control

Some reflexes can be modified by conscious control. You have picked up a hot plate again. But this time you are very hungry and your dinner is on the plate. This time your brain controls the reflex to stop you dropping your food.

Fill in the flow diagram to describe the nerve impulses for this outcome.

| you pick up a very hot plate | | |

| you hold on to the plate until you can put it down safely | |

---

## (1) Learning is the result of experience

Learning creates pathways in the brain that are more likely to transmit impulses than others.

**a** Draw a straight line to match each key word to its correct meaning.

| adapting | knowledge or skills gained from experience |
|---|---|
| learning | doing the same thing more than once |
| neuron pathways | adjusting to new conditions |
| repetition | different routes that electrical impulses can take through the brain |

**b** Human babies learn very quickly. Number these sentences to explain the sequence of events in the brain during learning. One has been done for you.

[1] The cortex in a baby's brain is a complicated network of neurons.

[ ] The response is learned.

[ ] Strengthened connections make it easier for more impulses to travel along the pathway.

[ ] A new experience sets up new pathways between the neurons in the cortex.

[ ] Using the pathway strengthens the connections between the neurons.

**c** Explain why repetition helps you learn a new sporting or musical skill.

.......................................................................................................................................

.......................................................................................................................................

.......................................................................................................................................

**d** Use these words to complete the sentences about brain development.

| knowledge   repetition   feral   older   difficult   easy   lost   gained |

As the brain develops, experience and _____ strengthen some

neural pathways. Some connections that have not been repeated at a later

stage are _____. There is evidence from studying _____ children

that children may only acquire some skills at a certain age. Children

neglected in early childhood find learning language _____.

## ② Learning new skills

An animal can adapt well to new situations if it has a variety of potential pathways in the brain.

**a** Adapting to new situations means learning new skills. Explain why humans are good at learning new skills throughout their life.

.......................................................................................................................................

.......................................................................................................................................

.......................................................................................................................................

**b** As we get older it becomes harder for the language-processing area in the cortex to make new pathways. Explain how this would affect an adult learning a new language.

.......................................................................................................................................

.......................................................................................................................................

**c** Describe how you could strengthen new pathways in the language-processing area of the cortex when learning a new language.

...........................................................................................................................

**d** Use these words to complete the sentences about brain development.

| knowledge repetition feral older difficult easy lost gained |
| --- |

As the brain develops, experience and ........................................ strengthen

some neural pathways. Some connections that have not been repeated at a

later stage are ................................... . There is evidence from

studying ................................... children that children may only acquire

some skills at a certain age. Children neglected in early childhood

find learning language ................................... .

**H**   **What is memory?**

## (1) Short-term and long-term memory

Memory can be long term or short term.

**a** Highlight or <u>underline</u> two phrases that together describe memory.

| learnt behaviour | storage of information | processing of information |
| --- | --- | --- |
| retrieval of information | input of sensory information | |

**b** Complete the sentence.

Looking at an unfamiliar phone number just before you dial it is an

example of ................................... memory.

**c** Verbal memory can be divided into short-term memory and long-term memory. Complete the table comparing these.

| Memory type | How long does it last? | How much can be stored? | An example |
| --- | --- | --- | --- |
| short-term | | | remembering that this row is about short-term memory |
| long-term | | | remembering the date of your birthday |

**d** Short-term and long-term memory work separately in the brain. Describe some evidence for this.

...........................................................................................................................

...........................................................................................................................

...........................................................................................................................

2 Kevin is a street performer. He can memorise a complete deck of shuffled cards in less than one minute.

a Describe three ways in Kevin can remember this large amount of information.

_____

_____

_____

_____

b Use the techniques described in part a to remember this hand of playing cards.

Write down the technique that you used.

_____

_____

_____

_____

3 **Scientists use models to describe how memory works**

a Draw a diagram of the multistore memory model in the space below.

b Explain why models are limited in explaining how memory works.

_____

_____

_____

### ① Sectors in the chemical industry

This chart shows the value of the sales of the various sectors of the chemical industry in the EU.

The figures are for 2004.

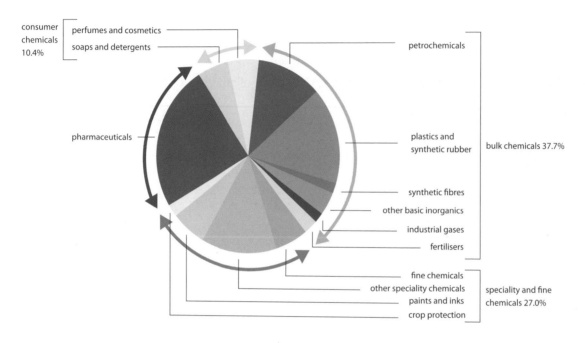

**a** What is the raw material for making petrochemicals? ............................................

**b** Give an example of a petrochemical that is used to make plastics on a large scale. ...............

**c** Give an example of a type of chemical needed for 'crop protection'. ...............................

**d** Why are paints and inks so valuable as products? ...............................................

..........................................................................................................

**e** Give the chemical formulae of the molecules of these three gases that are used in industry.

   **i** Nitrogen ..........................................

   **ii** Oxygen ..........................................

   **iii** Hydrogen ..........................................

**f** What percentage value of products of the EU chemical industry is used for the following?

   **i** Perfumes, cosmetics, soaps, and detergents ..........................................

   **ii** Pharmaceuticals ..........................................

## Acids and alkalis

① **Properties of acids and alkalis**

**a** Complete this table about some acids.

| Name of acidic compound | Formula of acidic compound | State of the pure compound at room temperature |
|---|---|---|
| citric acid | $C_6H_8O_7$ | |
| tartaric acid | $C_4H_6O_6$ | |
| | $H_2SO_4$ | |
| | $HNO_3$ | |
| ethanoic acid | $CH_3COOH$ | |
| hydrogen chloride | | |

**b** Complete this table about some alkalis.

| Name of alkaline compound | Formula of alkaline compound | State of the pure compound at room temperature |
|---|---|---|
| | NaOH | |
| potassium hydroxide | | |
| calcium hydroxide | | solid |

**c** Acids and alkalis show their characteristic reactions when dissolved in water.
Draw a line to match each of these solutions to its pH value.

| | |
|---|---|
| pure water | pH 14 |
| dilute hydrochloric acid | pH 12 |
| dilute sodium hydroxide solution | pH 7 |
| vinegar | pH 1 |
| limewater (calcium hydroxide solution) | pH 3 |

## ② Reactions of acids

- Complete the general word equations to summarise the main reactions of solutions of acids in water.
- Next complete the word equations for the examples.
- Finally complete and balance the matching symbol equations, including state symbols.

### a Acids with metals

The general word equation for the reaction of an acid with a metal is:

acid + _____ → salt + hydrogen

Example

_____ + magnesium → magnesium chloride + hydrogen

$HCl($___$)$  +  ___$($___$)$ →  $MgCl_2($___$)$  +  ___$($___$)$

### b Acids with metal oxides

The general word equation for the reaction of an acid with a metal oxide is:

acid + metal oxide → salt + _____

Example

nitric acid  + copper oxide  →  _____ + _____

___$($___$)$ +  $CuO($___$)$  →  $Cu(NO_3)_2($___$)$  +  ___$($___$)$

### c Acids with metal hydroxides

The general word equation for the reaction of an acid with a metal hydroxide is:

acid + metal oxide → _____ + water

Example

_____ + _____ → sodium sulfate  + _____ .

$H_2SO_4($___$)$  +  $NaOH($___$)$  →  ___$($___$)$ + $H_2O($___$)$

### d Acids with metal carbonates

The general word equation for the reaction of an acid with a metal carbonate is:

acid + metal carbonate → _____ + carbon dioxide + water

Example

_____ + _____ → calcium chloride + carbon dioxide + _____

$HCl($___$)$  +  $CaCO_3($___$)$  →  ___$($___$)$ + ___$($___$)$ + $H_2O($___$)$

## Salts from acids

### ① Acids, alkalis, and salts

**a** Complete the table below to show the products when some acids and alkalis react.

| Reaction | Alkali | Acid | Salt |
|---|---|---|---|
| 1 | sodium hydroxide | nitric acid | |
| 2 | potassium hydroxide | citric acid | |
| 3 | lithium hydroxide | hydrochloric acid | |
| 4 | sodium hydroxide | sulfuric acid | |
| 5 | lithium hydroxide | tartaric acid | |

**b** Write down the formulae of the salts produced in reactions 1, 3, and 4.

........................................................................................................................................................

........................................................................................................................................................

........................................................................................................................................................

**c** Draw a line to match each statement on the left to the related box on the right.

| | |
|---|---|
| the salt formed when sodium hydroxide reacts with sulfuric acid | NaOH |
| the salt formed when potassium hydroxide reacts with citric acid | soluble metal hydroxides |
| the acid that reacts with calcium hydroxide to form calcium nitrate | neutralisation |
| the alkali that reacts with acetic acid to form sodium acetate | sodium sulfate |
| a set of compounds that are alkalis in water | $HNO_3$ |
| the type of reaction that occurs when an acid and an alkali form a salt | potassium citrate |

## ② Ions and formulae

The table shows the charges on common ions found in ionic compounds.

**a** Complete the table.

| Positive ions | | | Negative ions | | |
|---|---|---|---|---|---|
| Ion | Charge | Symbol | Ion | Charge | Symbol |
| lithium | ............ | $Li^+$ | chloride | $1^-$ | ............ |
| sodium | $1+$ | ............ | ............ | $1^-$ | $Br^-$ |
| | $1+$ | $K^+$ | iodide | ............ | $I^-$ |
| | | | nitrate | $1^-$ | $NO_3^-$ |
| magnesium | $2+$ | ............ | ............ | $1^-$ | $OH^-$ |
| ............ | ............ | $Ca^{2+}$ | oxide | $2^-$ | ............ |
| barium | $2+$ | $Ba^{2+}$ | ............ | ............ | $CO_3^{2-}$ |
| ............ | $3+$ | $Al^{3+}$ | sulfate | ............ | $SO_4^{2-}$ |

**b** Use the complete table to help you to write down the formulae of these ionic compounds.

sodium hydroxide .......................................    magnesium carbonate ......................................

sodium chloride .......................................    magnesium sulfate ......................................

sodium carbonate .......................................    calcium carbonate ......................................

potassium nitrate .......................................    calcium chloride ......................................

potassium chloride .......................................    calcium iodide ......................................

magnesium bromide .......................................    calcium nitrate ......................................

magnesium oxide .......................................    aluminium oxide ......................................

magnesium hydroxide .......................................    aluminium chloride ......................................

**c** Work out the symbol for the ions in these compounds that are not included in the table.

   **i** Sulfide ion in barium sulfide, BaS ...........................................

   **ii** Strontium ion in strontium nitrate, $Sr(NO_3)_2$ ...........................

   **iii** Phosphate ion in sodium phosphate, $Na_3PO_4$ ...........................

## ③ Ionic theory of neutralisation

Use these words and symbols to complete the text.

| | | | | | | |
|---|---|---|---|---|---|---|
| acid | alkali | hydrogen | hydroxide | ions | molecules | salt |
| sodium | $H^+$(aq) | $H_2O$(l) | $H_2O$(l) | $K^+$(aq) | $NO_3^-$(aq) | $OH^-$(aq) |

- Acids are chemicals containing _____ atoms which react with

  water to give hydrogen _____ in solution..

  $HNO_3$(l) $\xrightarrow{water}$ _____ + $NO_3^-$(aq)

- Alkalis are ionic compounds. Examples are the soluble hydroxides of the alkali

  metals (lithium, _____, and potassium). These compounds

  consist of metal ions and _____ ions. When they dissolve

  they add hydroxide ions to water.

  KOH(s) $\xrightarrow{water}$ $K^+$(aq) + _____

- Potassium hydroxide and nitric acid react to make a _____
  (potassium nitrate) and water.

  _____ + $OH^-$(aq) + $H^+$(aq) + _____ → $K^+$(aq) + $NO_3^-$(aq) + _____

- During a neutralisation reaction the hydrogen ions from the _____

  react with the hydroxide ions from the _____ to make water

  $H^+$(aq) + $OH^-$(aq) → _____

  The remaining ions in solution make a salt.

## Purity of chemicals

## ① Testing the purity of citric acid

**Procedure**

Step 1: A 0.48 g sample of citric acid was dissolved in 50 cm³ of water.

Step 2: 1 cm³ phenolphthalein indicator was added.

Step 3: The solution was titrated with a solution of sodium hydroxide. This was
repeated a further three times. These were the burette readings from the four titrations.

| Titration | 1 | 2 | 3 | 4 |
|---|---|---|---|---|
| 2nd burette reading (cm³) | 22.40 | 19.40 | 24.40 | 25.90 |
| 1st burette reading (cm³) | 3.30 | 0 | 5.60 | 1.20 |
| Volume of NaOH(aq) added (cm³) | | 19.40 | | |

**a** Complete the table to show the volume of sodium hydroxide solution added in each titration.

**b** Shade in the box of any result that you think is an outlier.

**c** You decide to discard the outlier because you believe it was caused by an error in identifying the end point.

Find the mean of the remaining results for the volume of sodium hydroxide solution added.

........................................................................................................................................................

........................................................................................................................................................

**d** Why is it important to carry out repeat measurements?

........................................................................................................................................................

........................................................................................................................................................

**e** Label these diagrams to summarise the procedure for the titration.

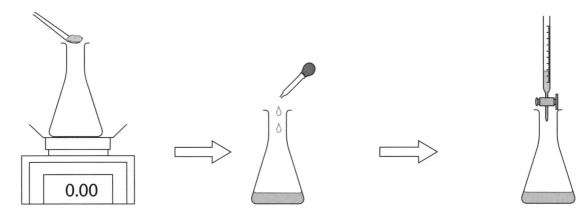

**f** Calculate the purity of the citric acid from the formula given below.
*T* is the titre (the volume of solution added from the burette).
The concentration of the sodium hydroxide used in the titration means that the value of *F* = 0.025.

$$\text{purity} = \frac{T \times F \times 100}{\text{mass of sample}} \, \%$$        purity = ................................. %

**g** Suggest when it is important to be able to measure the purity of a chemical such as citric acid.

........................................................................................................................................................

........................................................................................................................................................

## Energy changes in chemical reactions

### ① Endothermic and exothermic reactions

A student carried out an experiment. She mixed some chemicals together
and measured the temperature at the start and the end.

**a** Complete the following table to show the change in temperature. Then indicate
if the reactions were endothermic or exothermic.

| | Chemicals mixed | Temp at start (°C) | Temp at end (°C) | Change in temp (°C) | Endothermic or exothermic? |
|---|---|---|---|---|---|
| A | anhydrous copper sulfate and water | 20 | 28 | | |
| B | hydrochloric acid and zinc | 19 | 25 | | |
| C | citric acid, sodium hydrogencarbonate, and water | 18 | 10 | | |
| D | water and potassium chloride | 19 | 17 | | |
| E | ethanoic acid and sodium carbonate | 19 | 16 | | |
| F | calcium and water | 17 | 29 | | |
| G | magnesium and lead nitrate solution | 18 | 35 | | |

**b** The diagrams below represent energy changes in a chemical reaction.
Label each of them as *endothermic* or *exothermic* by putting a ring
around the correct **bold** word.

This graph shows an **exothermic /endothermic** reaction.

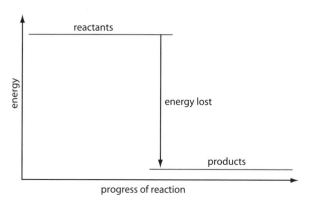

This graph shows an **exothermic /endothermic** reaction.

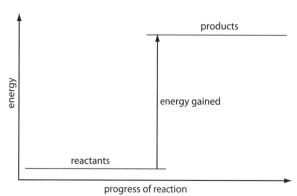

**c** Write the letters of the reactions from the table above next to the graph
that best represents them.

**d** Complete the sentences below by putting a ring around the correct **bold** word.

In an **exothermic / endothermic** reaction the temperature rises. The reaction
**gives / takes** energy **in from / out to** the surroundings.

In an **exothermic / endothermic** reaction the temperature decreases.
The reaction **gives / takes** energy **in from / out to** the surroundings.

## ① Effect of surface area on the rate of reaction

The diagrams show the apparatus for investigating the rate of reaction of marble chips (calcium carbonate) with acid.
The balanced equation for the reaction is:

$$CaCO_3(s) + 2HCl(aq) \rightarrow CaCl_2(aq) + CO_2(g) + H_2O(l)$$

- Diagram A shows the apparatus before the reaction starts.

- Diagram B shows the reaction in progress.

The reaction was carried out twice – first with larger chips of marble, then with smaller chips. There were still unchanged marble chips in both flasks when the reaction stopped.

The graph shows typical results using two sizes of marble chips.

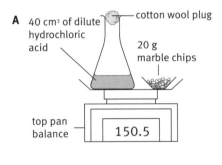

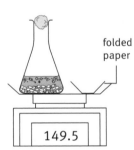

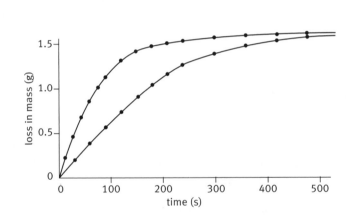

**a** Label the graph lines to show:

    **i** which is the line for larger chips and which the line for smaller chips

    **ii** where the reaction was fastest

    **iii** where the reaction was slowing down

    **iv** where the reaction had stopped.

**b** Explain why the total mass of the flask, acid, and marble fell during the reaction.

 

**c** Explain why the reaction slowed down and stopped with the same final loss in mass for both the larger chips and the smaller chips.

 

**d** Explain the difference in the rate of reaction at the start for the two sizes of marble chips.

## ② **Effect of concentration on the rate of reaction**

Adding dilute hydrochloric acid to a solution of sodium thiosulfate starts a slow reaction.

$$Na_2S_2O_3(aq) + 2HCl(aq) \rightarrow 2NaCl(aq) + H_2O(l) + SO_2(aq) + S(s)$$

The mixture turns cloudy. In time it is not possible to see through the solution.

The diagram shows a method for investigating the rate of reaction. The experimenter looks down at the cross on the paper from above and records the time it takes for the cross to become invisible after adding the acid.

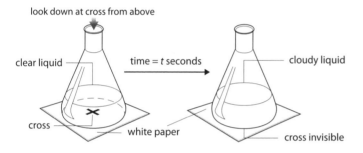

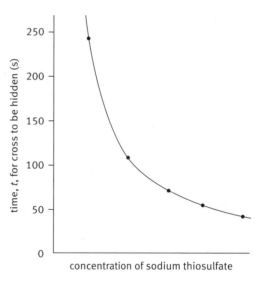

In an investigation, an experimenter added 5 cm³ of dilute hydrochloric acid to 50 cm³ samples of sodium thiosulfate solution. The experimenter measured the time for the cross to disappear with five different concentrations of sodium thiosulfate solution. The graph is a plot of the results.

**a** Look at the equation above and use it to explain why the solution of sodium thiosulfate turned cloudy after the hydrochloric acid was added.

**b** Why did the experimenter use the same volume and concentration of dilute hydrochloric acid with each different concentration of sodium thiosulfate solution?

**c** Put into words what the graph shows about the effect of changing the concentration of the sodium thiosulfate solution on the rate of the reaction.

## ③ The effect of temperature changes on the rate of reaction

The reaction of sodium thiosulfate with hydrochloric acid (see Question 2) can also be used to investigate the effect of temperature on the rate of a reaction. In this investigation, the volume and concentration of the acid and sodium thiosulfate stay the same. The thiosulfate solution is warmed before adding the acid. The experimenter measures the temperature of the mixture after adding the acid and the time taken for the cross to disappear.

**a** Why is it important to warm the sodium thiosulfate *before* adding the acid?

........................................................................................................................

........................................................................................................................

**b** Which factor is being *changed* in this investigation?

........................................................................................................................

**c** Why is it important to keep all the other possible factors the same during the investigation?

........................................................................................................................

........................................................................................................................

## ④ Predicting rates of reaction

The graph shows the volume of hydrogen produced when excess zinc granules react with 50 cm³ dilute hydrochloric acid at 20 °C, plotted against time.

Show the effect of each change to the conditions by completing this table and also by drawing and labelling on the graph the line you would expect.

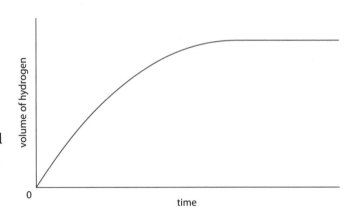

| Change of conditions (all other factors stay the same) | Effect on the rate of reaction at the start | Effect on the volume of gas collected (at room temperature) when the reaction stops |
|---|---|---|
| halving the concentration of the acid | | |
| carrying out the reaction at 30 °C | | |
| using the same mass of zinc but in larger pieces | | |

# 5) The effect of catalysts on rates of reaction

Hydrogen peroxide solution contains the compound $H_2O_2$.
At room temperature it decomposes very slowly to give water and oxygen.

$$2H_2O_2(aq) \longrightarrow 2H_2O(l) + O_2(g)$$

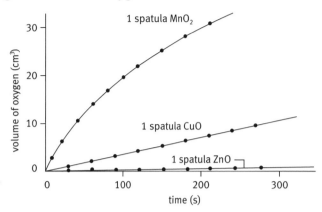

The graph shows the results from three tests.
Each time a small amount of a metal oxide was
added to 50 cm³ of a solution of hydrogen peroxide
in a flask. The oxygen gas was collected and
measured for up to five minutes.

In a control experiment, with no added metal oxide,
no oxygen was collected in five minutes.

**a** Complete and label this diagram to show how
the oxygen could be collected and measured.

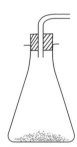

**b** Put the three metal oxides tested in order of their effectiveness as catalysts
for the reaction, with the least effective first.

_____

**c** Why was a control experiment carried out?

_____

**d** Explain the meaning of the term 'catalyst'.

_____

_____

# 6) Controlling reaction rates

These factors affect the rates of reactions:

- changes in concentration
- changes in temperature
- changes in the surface area of solids
- adding catalysts

Show how the rate of a reaction can be controlled by describing briefly how these reactions
can be slowed down.

**a** Milk going sour _____

_____

**b** Iron rusting in air _____

_____

# ⑦ Collision theory

Chemists use collision theory to explain why factors such as surface area
and concentration affect the rates of reactions.

**a** Write a paragraph of four or five sentences to explain the key ideas of collision theory.

_____

_____

_____

_____

_____

_____

_____

**b** Colour and label this diagram. Then use it in an explanation to show how
collision theory explains why reactions in solution go faster if the
concentrations of reactants are higher.

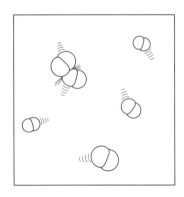  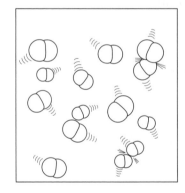

Explanation _____

_____

_____

_____

## Stages in chemical synthesis

### ① Making a soluble salt from an acid

The diagrams show how to make a pure sample of magnesium sulfate from magnesium oxide.

**a** Label the diagrams. Add a caption to each stage of the diagram to describe what is happening and to explain the purpose of the stage.

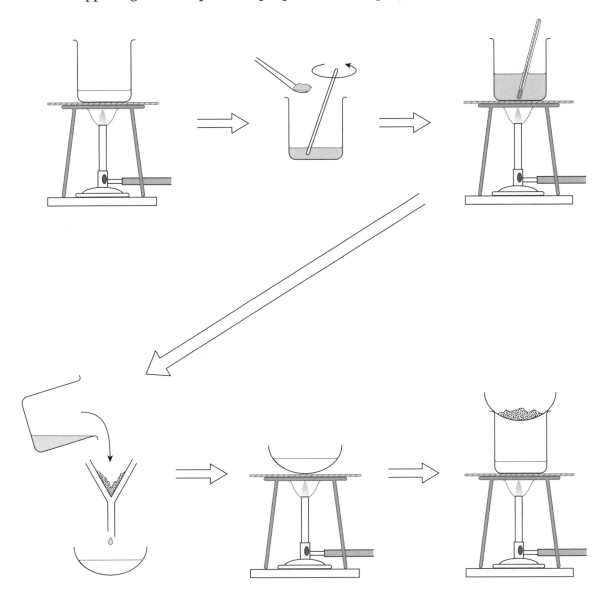

**b** Write a balanced chemical equation for the reaction above.

## ② Stages in synthesis

Use the example on the previous page (or any other example) to explain the importance of these stages in any chemical synthesis.

**a** Choosing the reaction

..........................................................................................................................................................

..........................................................................................................................................................

..........................................................................................................................................................

**b** Working out the quantities to use

..........................................................................................................................................................

..........................................................................................................................................................

..........................................................................................................................................................

**c** Carrying out the reaction in suitable apparatus under the right conditions

..........................................................................................................................................................

..........................................................................................................................................................

..........................................................................................................................................................

**d** Separating the product from the reaction mixture

..........................................................................................................................................................

..........................................................................................................................................................

..........................................................................................................................................................

**e** Purifying the product

..........................................................................................................................................................

..........................................................................................................................................................

..........................................................................................................................................................

**f** Measuring the yield and checking the purity of the product

..........................................................................................................................................................

..........................................................................................................................................................

..........................................................................................................................................................

## Chemical quantities

### 1 Yields

**a** Work out the theoretical yield of magnesium sulfate, $MgSO_4$, that can be made from 4.0 g of magnesium oxide, MgO, and an excess of sulfuric acid.

- **Step 1** Write the balanced symbol equation for the reaction. (Write it on the top line only.)

- **Step 2** Work out the formula masses of the chemicals. Relative atomic masses: Mg = 24, O = 16, S = 32, H = 1

- **Step 3** Write the relative reacting masses for the relevant chemicals under the balanced equation in Step 1.

- **Step 4** Convert to reacting masses by writing in the units.

- **Step 5** Scale the quantities to the amounts actually used to find the theoretical yield.

**b** What is the percentage yield if the actual yield is 10.0 g?

**c** Use the same steps to work out the theoretical yield of zinc sulfate, $ZnSO_4$, that can be made from 9.2 g zinc carbonate, $ZnCO_3$, and an excess of sulfuric acid. (Relative atomic masses: Zn = 65, C = 12, S = 32, O = 16)

**d** What is the percentage yield if the actual yield is 11.4 g?

## ① Changes inside the atom – radioactive decay

Carbon has more than one type of atom. In a sample of carbon, most of the atoms will be carbon-12 atoms, which are stable. However, a tiny proportion will be carbon-11 atoms. These are radioactive.

**a** The materials on the left all contain carbon. Draw lines to link each material with *four* of the statements on the right. Each statement can be used once, more than once, or not at all. The first line has been drawn for you.

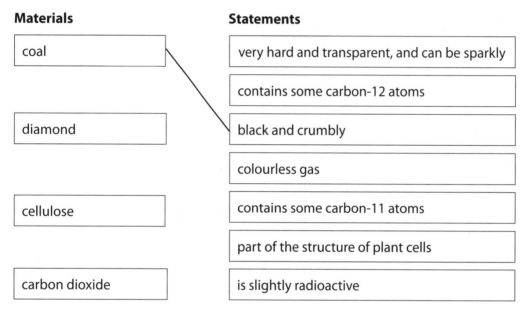

**Materials**

- coal
- diamond
- cellulose
- carbon dioxide

**Statements**

- very hard and transparent, and can be sparkly
- contains some carbon-12 atoms
- black and crumbly
- colourless gas
- contains some carbon-11 atoms
- part of the structure of plant cells
- is slightly radioactive

**b** Three of the statements apply to all of the materials – they do not depend on the chemical properties of the material. Put a tick ✓ next to these statements.

**c** Complete these sentences. Draw a ring around the correct **bold** words.

The materials above have **the same / different** chemical properties. However, they all contain atoms of the **elements / compound** carbon. Some of these atoms are **radiological / radioactive**. The radioactive carbon **is / is not** affected by the chemical properties of the materials. A carbon-11 atom is radioactive because its **nucleus / neighbour** is **stable / unstable**. It is stability of the **nucleus / lattice** that determines the stability of the atom.

## ② Radioactive emissions

Join the boxes to show what the types of radiation are.

**What it is**

- alpha particles (α)
- beta particles (β)
- gamma rays (γ)

**Radiation**

- very small, very high-speed particles with negative charge
- high-energy electromagnetic radiation
- small, high speed particles with positive charge

## Atoms and nuclei

① **Alpha particle scattering**

This diagram shows a view from above of the apparatus used in Rutherford's alpha-scattering experiment.

**a** Label the diagram by writing these words in the correct boxes.

| | | |
|---|---|---|
| alpha source | angle of deflection | beam of alpha particles |
| movable microscope | screen | thin gold foil | vacuum |

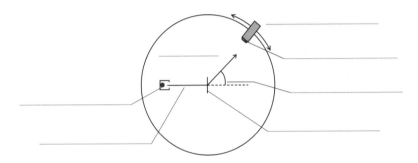

**b** Why it is essential that the experiment is carried out in a vacuum chamber?

......................................................................................................................

**c** Alpha radiation is directed at a sheet of gold foil. The diagram below shows what was observed. Describe two important observations from the experiment.

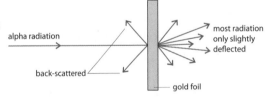

Observations:

1 ..................................................................................................................

......................................................................................................................

2 ..................................................................................................................

......................................................................................................................

**d** This table shows Rutherford's conclusions. Fill in the reasons for his conclusions.

| Conclusion | Reason |
|---|---|
| There must be something in the gold with positive charge. | |
| The positive charge must have a lot of mass. | |
| The positive charge must be very small. | |

## ① **The structure of an atom**

Look at the diagram of an atom on the right.
The boxes describe parts of the diagram.

**a** Draw a line from each of these boxes to the correct part of the atom. The first one has been done for you. Each part can be linked to more than one box.

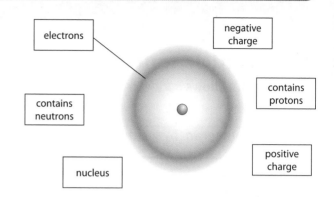

electrons

negative charge

contains protons

contains neutrons

positive charge

nucleus

**b** If the atom were drawn to the same scale as the nucleus, about what would its diameter be?

Draw a ring around your choice.

| 1 cm | 10 cm | 50 cm | 10 cm | 500 m |
|------|-------|-------|-------|-------|

**c** The nucleus contains two types of particle: the proton and the neutron. The statements on the left describe either the proton or the neutron. Link each statement to one of the boxes on the right. The first one has been done for you.

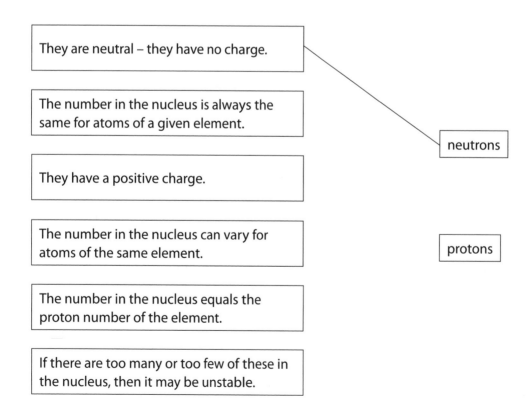

They are neutral – they have no charge.

The number in the nucleus is always the same for atoms of a given element.

They have a positive charge.

The number in the nucleus can vary for atoms of the same element.

The number in the nucleus equals the proton number of the element.

If there are too many or too few of these in the nucleus, then it may be unstable.

neutrons

protons

## ② Different atoms

The diagrams below represent four different atoms. The number of protons and neutrons is shown in the nucleus of each one.

**Z**

86 P
220 N

**a** Put a tick ✓ next to the two atoms that are from the same element.

**b** Atom Z is unstable and decays by giving out an alpha particle.

It changes into the atom shown on the right.

How can you tell that it is an atom of a different element?

_α particle_

**c** Draw an arrow from the atom to the alpha particle to show where the alpha particle came from.

## ③ Isotopes

**a** Use the words from the box to fill in the blanks in the sentences.

| chemical reactions | isotopes | mass | neutrons | protons |
|---|---|---|---|---|

Carbon-11 and carbon-12 are both ........................... of carbon. They have

the same number of ........................... and a different number of ........................... .

They will react the same way in ...........................

but they have different ........................... .

**b** This is the formula for describing the carbon-11 isotope: $^{11}_{6}C$

  **i** In the table below, draw one line from each box showing the **position of the number in the formula** to the box showing **what the number means**.

  **ii** Then draw one line from **what each number means** to the **value for the carbon-11 isotope**.

| Position of number in formula | What the number means | Number for carbon-11 |
|---|---|---|
| top | number of protons | $11 - 6 = 5$ |
| bottom | number of protons and neutrons | 6 |
| not in formula | number of neutrons | 11 |

## ④ Alpha and beta particles

In the sentences below, fill in the blanks to describe alpha (α) and beta (β) particles. Use words from the box once, more than once, or not at all. Draw the formula for each particle in the box provided.

| electron | electrons | four | nucleus | neutron |
|---|---|---|---|---|
| neutrons | one | proton | protons | three | two |

- An alpha particle is made of _____ _____ and _____ .

  It is the same as a helium nucleus. In nuclear equations it has the symbol:

  

- A beta particle is an _____ but it comes from the _____ .

  It is not an orbital _____ . It is formed when a _____

  decays to give a _____ and an _____ . In nuclear

  equations it has the symbol:

  

## ⑤ Nuclear equations

**a** Complete the equation to show the alpha decay.

Plutonium-239, used as a nuclear fuel and in nuclear weapons, decays to uranium.

$$^{239}_{94}\text{Pu} \longrightarrow \quad \overline{\quad}\text{U} + \quad \overline{\quad}\alpha$$

**b** Complete the equation to show the beta decay.

Nitrogen-16 is a radioactive isotope that decays to oxygen.

$$^{16}_{7}\text{N} \longrightarrow \quad \overline{\quad}\text{O} + \quad \overline{\quad}\beta$$

---

### D  Using radioactive isotopes

## ① Alpha, beta, and gamma radiation

**a** The types of radiation given off by radioactive materials have different properties. Complete the table by using ✓ and ✗ to record the different properties of the radiation.

| Type of radiation | Is absorbed by a thin sheet of paper | Is absorbed by a thin sheet of a aluminium | Is absorbed by a thick sheet of lead |
|---|---|---|---|
| alpha | | | |
| beta | | | |
| gamma | | | |

**b** Draw one line to link each **description** to the correct **radiation**, and another line to link the **radiation** to the correct **electric charge**.

| Description | Radiation | Electric charge |
|---|---|---|

most penetrating

most ionising

least ionising

has the most mass

fastest moving

alpha

beta

gamma

negative

neutral

positive

## 2 Radiation sources

Radioactive sources are tested by placing sheets of material between the source and a Geiger counter 2 cm away.

| Sheet added | Count rate (counts per second) | |
|---|---|---|
| | Source A | Source B |
| none | 7.5 | 6.8 |
| paper | 7.3 | 0.7 |
| 3-mm-thick | 5.9 | 0.9 |
| 3-cm-thick lead | 0.9 | 0.8 |

**a  i** Which type or types of radiation does source A emit? _____

**ii** Explain your reasoning.

_____

_____

**b  i** Which type or types of radiation does source B emit? _____

**ii** Explain your reasoning.

_____

_____

## 3 Sterilisation

Give two reasons why gamma radiation is used to sterilise disposable syringes.

1 _____

2 _____

## ① **Radiation dose from different sources**

You are continually exposed to background radiation. Radiation dose is a measure of the possible harm to your body's cells caused by radiation.

Josy calculated the radiation dose she received last year. Her estimates (in millisieverts, mSv) are shown in the table below.

| Source of radiation | Josy's dose (mSv) | Natural or artificial? | Occupations with increased dose |
|---|---|---|---|
| radon in air where she lives | 0.700 | | |
| from rocks and buildings | 0.030 | | |
| cosmic rays where she lives | 0.230 | | |
| cosmic rays from air travel | 0.020 | | |
| from food and drink | 0.300 | | |
| from medical treatments | 0.040 | | |
| from nuclear industry and fallout | 0.017 | | |
| Total | **1.337** | | |

**a** Write either **N** or **A** in the third column to show which sources are natural and which are artificial.

**b** What is Josy's total dose from:

   **i** natural sources?........................................ mSv

   **ii** artificial sources? ........................................ mSv

**c** What effect do the artificial sources have on the risk to Josy from radiation? Draw a ring around your choice.

| a large increase     a small increase     no effect     a small decrease |
|---|

**d** The national average radiation dose is 2.600 mSv for one year. Compared with the national average, suggest the size of the risk to Josy from radiation. Draw a ring around your choice.

| more than the national average     less than the national average     much less than the national average |
|---|

Large radiation doses can cause cancer. People exposed to 1000 mSv have a 3% chance of getting cancer due to the radiation.

**e** How does the chance of Josy getting cancer from radiation compare with this? Draw a ring around your choice.

| more than 3%     about 3%     less than 3%     much less than 3% |
|---|

**f** Some people have occupations that increase their radiation dose. Choose three sources of radiation dose that could be increased by a person's job. Write the job in the last column of the table (on the previous page).

## ② **Irradiation or contamination**

You are exposed to radiation in the environment by irradiation and by contamination.

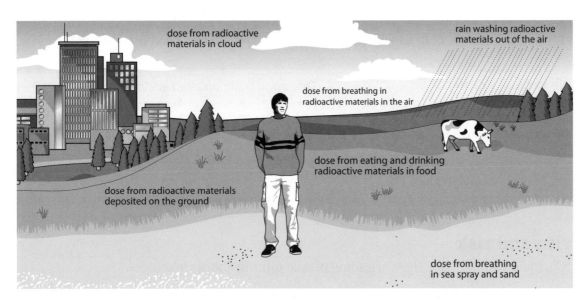

The picture describes some of the sources of radiation you are exposed to.
Next to each description, write:

*   **I** if you think it is an example of **i**rradiation

*   **C** if you think it is an example of **c**ontamination

## ③ **Radiation and health – protecting patients and staff**

**a** Look at these occupations. Put a tick ✓ next to those people who are 'radiation workers'.

☐ a hospital radiologist ☐ a chef in a 'bar and grill'

☐ a scientist in a nuclear power station ☐ a technician who sterilises surgical

☐ a radio broadcaster ☐ equipment with gamma rays

**b** Radiologists regularly work with radioactive sources. It is important to take precautions to minimise the risk from the radiation. Look at the precautions below. Draw a line to match each **precaution** to the **reason** for it.

| Precaution | Reason |
|---|---|
| wear gloves and an apron | to block out the radiation and reduce the dose and risk due to irradiation |
| wear protective clothing and stand behind a screen | to monitor the radiation dose |
| wear a special badge that is sensitive to ionising radiation | to prevent clothes and skin being contaminated with sources of ionising radiation |

## ① **Radon gas**

Radon is a gas that is produced naturally in some rocks. It emits alpha radiation.

The statements below describe why radon is hazardous because of contamination. But they are in the wrong order.

Draw arrows to link the statements in the correct order. The first one has been done for you.

| |
|---|
| damage the cells or cause a cancer. |

| |
|---|
| are absorbed by soft, internal tissue. They can |

| |
|---|
| breathe in atoms of the gas, which might |

| |
|---|
| The radon build up in an enclosed area. People |

| |
|---|
| ionise atoms in cells; this can |

| |
|---|
| give off alpha radiation inside a person's lungs. The alpha particles |

## ② **Radioactivity and risk**

**a** There are risks in all walks of life. Here are some people talking about risks.

Draw a line to match each statement to the best explanation.

**Statements about risk** | **Explanations**

**radiologist** — My maximum-allowed dose is 20 mSv per year, but it is kept well below this level at about 1.5 mSv per year.

This is an example of a new technology presenting a new risk.

**airline pilot** — It is true that air travel has exposed us to new risks, such as cosmic radiation and deep vein thrombosis (DVT), but people still want to fly.

This is because the risk depends on the total radiation dose you have received.

**doctor** — The risk to your health of not doing the investigation using radiation is greater than the risk from the radiation itself.

This is an example of trying to keep the risk as low as possible.

**patient** — The doctor wanted to know how many X-ray investigations and radiotherapy treatments I had had in the past.

Because radiation is invisible some people may think the risk is greater than it is.

**recruitment officer for radiologist** — Some people are surprised to find out that the risk of death per year for construction workers (1 in 16 000) is similar to that for radiation workers (1 in 17 000).

This is an example of balancing the benefits againts the risks.

**b** Look at the statement by the recruitment officer. (Ring) the correct **bold** words to complete these sentences:

**i** For a construction worker the perceived risk is **greater** / **less than** the actual risk.

**ii** For a radiation worker the actual risk is **greater** / **less than** the perceived risk.

# Half-life

## The pattern of radioactive decay

**a** Complete these sentences. Draw a ring around the correct **bold** words.

The amount of radiation from a radioactive material is called its
**activity / amplitude**. This **decreases / increases** over time. The time it takes
for the activity to drop **by a half / to zero** is called the half-life. Different
radioactive sources **can / do not** have very different half-lives. The
**longer / shorter** the half-life, the longer a source will be radioactive.

**b** Two radioactive isotopes and their half-lives are given in the table.

| Radioactive isotopes | Iodine-131  (A) | Iodine-129  (B) |
|---|---|---|
| Half-life | 8 days | 15.7 million years |

Each of these graphs has three points marked on the curve, showing the
amount of radioactive decay at different times. For each graph, complete the
times on the horizontal axis.

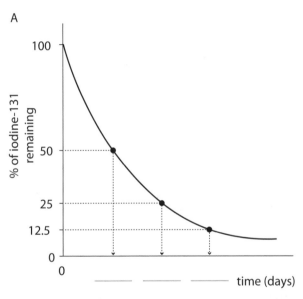

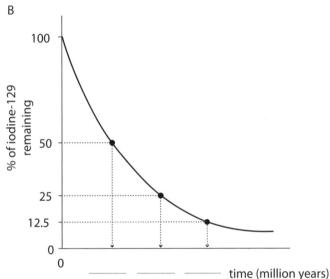

**c** A sample contains 64 g of iodine-131.

**i** How much iodine-131 will be left behind after each of these time intervals?

- 8 days ................... g
- 16 days ................... g
- 24 days ................... g

**ii** How long will it take until the amount of iodine-131 has dropped to 2 g?

number of half-lives = ................... so time = ................... days

**iii** Tick ✓ the correct phrase below to complete the sentence.

The activity of a sample is low enough to be considered safe:

☐ when the activity has fallen to one tenth
of its starting value

☐ after one tenth of the half-life

☐ after ten half-lives

☐ when the activity is similar to background
radiation

## ② Radioactive decay curve

This is the radioactive decay curve for a sample of carbon-14.

It starts with an activity of 60 counts per minute.

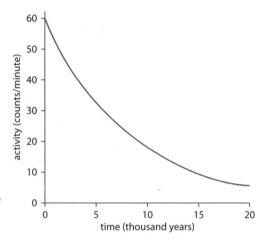

**a** What will the activity be after one half-life?

............................................................................ counts/min

**b** Draw a horizontal line on the graph to show this.

**c** Draw a vertical line to the x-axis to show what the half-life is, and label this value 'half-life'.

**d** Draw a ring around the value of the half-life.

| 6 years | 10 years | 30 years | 6000 years | 10 000 years | 30 000 years |

## ③ A sample of technetium

This is the radioactive decay curve for a sample of technetium-99m.

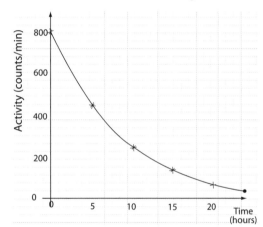

| Time (h) | Activity (counts/min) |
|----------|----------------------|
| 0 | 800 |
| 5 | 440 |
| 10 | 250 |
| 15 | 140 |
| 20 | 80 |

**a** What is the initial activity of the sample in counts/min? ......................... counts/min

**b** Draw a horizontal and a vertical line on the graph to show the half-life.

**c** What is the half-life in hours? ......................................... h

**d** Use your answer to part **a** to work out the activity after two half-lives.

The activity after two half-lives is ............................... counts/min.

**e** Use your answer to part **c** to work out the time for two half-lives.

The time for two half-lives is ............................... hours.

**f** Plot on the graph the point given by your answers to parts **d** and **e**.

**g** Is the point you plotted on the decay curve?

## Medical imaging and treatment

### 1 Using ionising radiation

Ionising radiations are used in hospitals. Look at the types of radiation on the left.

**a** Draw lines from each box explaining how the dose is given to match it with the type of radiation in the middle column. There may be one or two lines from each box.

| **How dose given to patient** | **Type of radiation** | **Use** |
|---|---|---|
| passed through the body from an external source | beta radiation | providing images of organs to look for abnormal function |
| emitted by isotopes that are swallowed by, injected, or implanted in the patient | gamma radiation | treating cancers by killing the cancerous cells |

**b** Draw a line from each type of radiation to one or both of the uses on the right.

**c** Give two reasons why alpha radiation is not used to image internal organs.

1 _____ 2 _____

### 2 Benefits and risks

Complete the sentences below. Draw a ring around the correct **bold** words.

Some **medical / sporting** techniques use ionising radiations. Like any ionising radiation, they can damage **cells / equipment** and there is a chance that they will start a **cancer / virus**. However, the chance is very **small / large**. Most people think that the risk is outweighed by the **benefits / costs** of using the technique to investigate or treat a medical problem.

### 3 Balancing the cost

Jo lives in the UK. She has been feeling unwell and is given an injection of DMSA, which is taken up by her kidneys. It gives out gamma radiation and allows her doctors to get an image of the working parts of her kidneys. The gamma scan they are working normally. Joseph lives in Uzbekistan. He has been feeling ill. He is treated with painkillers.

**a** The gross domestic product (GDP) for the UK and Uzbekistan are €1800 and €900 per person per year.

Which country has the larger GDP? _____

**b** Suggest why is Joseph not given a gamma scan as a precaution.

_____

## (1) Nuclear fission

**a** Complete these sentences. Draw a ring around the correct **bold** words.

Uranium-235 is a **nuclear / fossil** fuel. If its nucleus absorbs a
**neutron / electron**, it becomes extremely **unstable / radioactive** and splits
into two. This is called nuclear **fission / fusion**. When the nucleus splits, it
**releases / absorbs** energy.

**b** The diagram shows how uranium-235 can 'chain react'.
The statements below describe the chain reaction. They are out
of sequence. Draw arrows to join the statements in the correct order.
The last statement will point back to one of the earlier statements
forming a loop. The first arrow has been drawn for you.

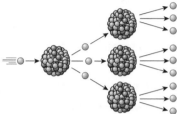

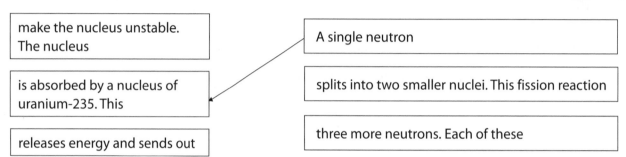

make the nucleus unstable.
The nucleus

A single neutron

is absorbed by a nucleus of
uranium-235. This

splits into two smaller nuclei. This fission reaction

releases energy and sends out

three more neutrons. Each of these

## (2) Nuclear reactor

**a** The diagram below shows the nuclear reactor and boiler parts of a nuclear
power station. Use the words and phrases from the box to label the diagram.

| boiler | concrete shield | control rods | coolant | fuel rods | nuclear reactor | steam | water |

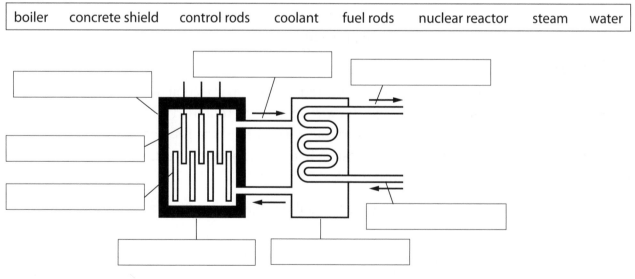

**b** Explain how the nuclear reactor uses heat from the fission reaction to produce
the steam for the turbines. Use all the labels from part **a** in your explanation.

.......................................................................................................................

.......................................................................................................................

.......................................................................................................................

# Energy from fission

Complete the sentences using these two words. You can use each word as many times as you like.

| fission | chemical |
|---|---|

- _____ is a nuclear reaction inside an atom, whereas combustion

is a _____ reaction between atoms.

- In combustion, _____ bonds are made and broken;

In _____ , a nucleus splits into smaller nuclei.

- A _____ reaction releases about a million times more

energy than a _____ reaction because the forces inside the

nucleus are so much bigger than the forces between atoms.

# Einstein's equation

The energy released from a nuclear reaction can be calculated using Einstein's famous equation $E=mc^2$.

$$\text{energy} \quad = \quad \text{mass} \quad \times (\text{speed of light in a vacuum})^2$$

$$(\underline{\hspace{3cm}}) = (\underline{\hspace{4cm}}) \times (\underline{\hspace{4cm}})^2$$

**a** Complete the equation by putting the correct units in the brackets.

**b** When a nucleus of uranium-235 absorbs a neutron it becomes a nucleus of uranium-236, which is very unstable.

**i** Complete this nuclear equation for the reaction.

$$^{235}_{92}U + \underline{\hspace{2cm}} n \longrightarrow \ ^{236} \quad U$$

**ii** The uranium-236 nucleus undergoes a fission reaction. It splits into two parts of similar size.
Complete this equation for the fission reaction.

$$^{236}_{92}U \longrightarrow \ ^{142}_{56}Ba + \ ^{91} \underline{\hspace{1.5cm}} Kr + 3^1_0 n$$

**iii** Draw a ring around the correct **bold** words to complete this explanation.

When the mass of all the fission products is added up it **equals** / **is less than** the mass of uranium-236 nucleus. The mass **difference** / **direction** has been released as **energy** / **electricity**. This **energy** / **electricity** can be calculated from $E = mc^2$.

## ① **Dealing with radioactive waste**

Decisions have to be made about how to deal with existing radioactive waste. This waste was produced by nuclear power stations and medical uses of radioactive materials.

**a** Draw a line to match each statement to the type of radioactive waste it best describes.

| Made up of the most dangerous fission products from used fuel rods. |  Low-level waste (LLW) | It is very bulky, packed in drums, and kept in specially protected landfill sites. |
| Made up of used protective clothing, and refuse and rubble with low radioactivity. |  Intermediate-level waste (ILW) | It is not very bulky but it gets so hot that it has to be stored under water. |
| Made up of materials from inside the reactor. It contains some fission products with very long half-lives. |  High-level waste (HLW) | It is chopped up and mixed with concrete, then stored in large steel containers. |

**b** Choose words from the box to complete the sentences below. Each word can be used once, more than once, or not at all.

| absorbed | food | hazardous | irradiation | safe |
| transmitted | contamination | | | |

Intermediate-level waste is _____ for a long time. The

radiation it gives off is _____ by its packaging so there is

very little risk from _____ . However, there is a long-term risk

from _____ . It must be kept out of water supplies and

the _____ chain for tens of thousands of years. Scientists are

still looking for a permanent disposal method that they are sure will

be _____ for these lengths of time. No permanent method will

be used until they are sure it is safe.

# Nuclear fusion

## 1) The strong nuclear force

**a** Draw a ring around the correct **bold** words to complete the statement.

Protons have a **negative** / **positive** charge. Two protons will **attract** / **repel** each other. The nucleus is made of **electrons** / **protons** and **beta particles** / **neutrons**. Neutrons have **charge** / **no charge**. This means that there must be another force to overcome the **attractive** / **repulsive** force and hold the nucleus together. This force is called the **strong** / **weak** nuclear force.

**b** Tick ✓ all the statements that are true.
The strong nuclear force:

☐ affects protons and neutrons

☐ affects protons and electrons

☐ affects protons, neutrons, and electrons

☐ is an attractive force between protons and neutrons

☐ is a repulsive force between neutrons

☐ is a repulsive force between protons

☐ is a very long range force

☐ is a very short range force

**c** The diagram shows the fusion of two hydrogen nuclei to create a helium nucleus.
Complete the diagram by filling in the missing numbers.

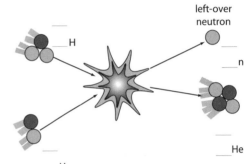

## 2) Nuclear fusion and fission

Join the **description** to the **reaction**. to name the reactions. Then show whether each reaction releases or stores energy.
Boxes may be used once, more than once, or not at all.

| Description | Reaction | Energy released or stored? |
|---|---|---|
| A large nucleus splits into two similar size nuclei. | fusion | Energy is released. |
| Two small nuclei join together to make one. | fission | Energy is taken in and stored in the nucleus or nuclei. |

## 3) Fusion research

**a** In an atom the positive nucleus is surrounded by a cloud of negative orbital electrons.
Give two reasons why very high temperatures are needed before nuclei can fuse.

1 .......................................................................................................................................

2 .......................................................................................................................................

**b** The ITER project is a joint fusion research project between many countries.

Give two advantages of scientists from different countries working together on large research projects.

1 .......................................................................................................................................

2 .......................................................................................................................................

# Appendices

## Useful relationships, units, and data

### Relationships

You will need to be able to carry out calculations using these mathematical relationships.

### P4 Explaining motion

$$\text{speed} = \frac{\text{distance}}{\text{time}}$$

$$\text{acceleration} = \frac{\text{change in velocity}}{\text{time taken}}$$

$$\text{momentum} = \text{mass} \times \text{velocity}$$

change of momentum = resultant force × time for which it acts

work done by a force = force × distance moved in the direction of the force

amount of energy transferred = work done

change in gravitational potential energy = weight × vertical height difference

kinetic energy = ½ × mass × velocity²

### P5 Electric circuits

power = voltage × current

$$\text{resistance} = \frac{\text{voltage}}{\text{current}}$$

$$\frac{\text{voltage across primary coil}}{\text{voltage across secondary coil}} = \frac{\text{number of turns in primary coil}}{\text{number of turns in secondary coil}}$$

### C6 Chemical synthesis

$$\text{percentage yield} = \frac{\text{actual yield}}{\text{theoretical yield}} \times 100\%$$

### P6 Radioactive materials

Einstein's equation: $E = mc^2$, where $E$ is the energy produced, $m$ is the mass lost, and $c$ is the speed of light in a vacuum.

## Units that might be used in the Additional Science course

length: metres (m), kilometres (km), centimetres (cm), millimetres (mm), micrometres (μm), nanometres (nm)

mass: kilograms (kg), grams (g), milligrams (mg)

time: seconds (s), milliseconds (ms)

temperature: degrees Celsius (°C)

area: cm², m²

volume: cm³, dm³, m³, litres (l), millilitres (ml)

speed and velocity: m/s, km/s, km/h

energy: joules (J), kilojoules (kJ), megajoules (MJ), kilowatt-hours (kWh), megawatt-hours (MWh)

electric current: amperes (A), milliamperes (mA)

potential difference/voltage: volts (V)

resistance: ohms (Ω)

power: watts (W), kilowatts (kW), megawatts (MW)

radiation dose: sieverts (Sv)

### Prefixes for units

| nano | micro | milli | kilo | mega | giga | tera |
|---|---|---|---|---|---|---|
| one thousand millionth | one millionth | one thousandth | × thousand | × million | × thousand million | × million million |
| 0.000000001 | 0.000001 | 0.001 | 1000 | 1000 000 | 1000 000 000 | 1000 000 000 000 |
| $10^{-9}$ | $10^{-6}$ | $10^{-3}$ | $\times 10^3$ | $\times 10^6$ | $\times 10^9$ | $\times 10^{12}$ |

# Useful information and data

## P4 Explaining motion

A mass of 1 kg has a weight of 10 N on the surface of the Earth.

## C5 Chemicals of the natural environment

Approximate proportions of the main gases in the atmosphere: 78% nitrogen, 21% oxygen, 1% argon, and 0.04 % carbon dioxide.

## P5 Electric circuits

mains supply voltage: 230 V

## P6 Radioactive materials

speed of light ($c$) = 300 000 000 m/s

# Chemical formulae

## C4 Chemical patterns

water $H_2O$, hydrogen $H_2$, chlorine $Cl_2$, bromine $Br_2$, iodine $I_2$

lithium chloride LiCl, lithium bromide LiBr, lithium iodide LiI

sodium chloride NaCl, sodium bromide NaCl, sodium iodide NaI

potassium chloride KCl, potassium bromide KBr, potassium iodide KI

lithium hydroxide LiOH, sodium hydroxide NaOH, potassium hydroxide KOH

## C5 Chemicals of the natural environment

nitrogen $N_2$, oxygen $O_2$, argon A, carbon dioxide $CO_2$

sodium chloride NaCl, magnesium chloride $MgCl_2$

sodium sulfate $Na_2SO_4$, magnesium sulfate $MgSO_4$

potassium chloride KCl, potassium bromide KBr

## C6 Chemical synthesis

chlorine $Cl_2$, hydrogen $H_2$, nitrogen $N_2$, oxygen $O_2$

hydrochloric acid HCl, nitric acid $HNO_3$, sulfuric acid $H_2SO_4$

sodium hydroxide NaOH, sodium chloride NaCl, sodium carbonate $Na_2CO_3$, sodium nitrate $NaNO_3$, sodium sulfate $Na_2SO_4$, potassium chloride KCl

magnesium oxide MgO, magnesium hydroxide $Mg(OH)_2$, magnesium carbonate $MgCO_3$, magnesium chloride $MgCl_2$, magnesium sulfate $MgSO_4$

calcium carbonate $CaCO_3$, calcium chloride $CaCl_2$, calcium sulfate $CaSO_4$

# Qualitative analysis data

## Tests for negatively charged ions

| Ion | Test | Observation |
| --- | --- | --- |
| carbonate $CO_3^{2-}$ | add dilute acid | effervesces, and carbon dioxide gas is produced (the gas turns lime water milky) |
| chloride (in solution) $Cl^-$ | acidify with dilute nitric acid, then add silver nitrate solution | white precipitate |
| bromide (in solution) $Br^-$ | acidify with dilute nitric acid, then add silver nitrate solution | cream precipitate |
| iodide (in solution) $I^-$ | acidify with dilute nitric acid, then add silver nitrate solution | yellow precipitate |
| sulfate (in solution) $SO_4^{2-}$ | acidify, then add barium chloride solution or barium nitrate solution | white precipitate |

## Tests for positively charged ions

| Ion | Test | Observation |
| --- | --- | --- |
| calcium $Ca^{2+}$ | add sodium hydroxide solution | white precipitate (insoluble in excess) |
| copper $Cu^{2+}$ | add sodium hydroxide solution | light-blue precipitate (insoluble in excess) |
| iron(II) $Fe^{2+}$ | add sodium hydroxide solution | green precipitate (insoluble in excess) |
| iron(III) $Fe^{3+}$ | add sodium hydroxide solution | red–brown precipitate (insoluble in excess) |
| zinc $Zn^{2+}$ | add sodium hydroxide solution | white precipitate (soluble in excess, giving a colourless solution) |

The Periodic Table of the Elements

Key:

```
    1
    H
 hydrogen
    1
```

- 1 — proton number
- H — symbol
- hydrogen — name
- 1 — atomic mass

| group number | 1 | 2 | | | | | | | | | | | | | 3 | 4 | 5 | 6 | 7 | 0 |
|---|---|---|---|---|---|---|---|---|---|---|---|---|---|---|---|---|---|---|---|---|
| **period 1** | 1 H hydrogen 1 | | | | | | | | | | | | | | | | | | | 4 He helium 2 |
| **period 2** | 7 Li lithium 3 | 9 Be beryllium 4 | | | | | | | | | | | | | 11 B boron 5 | 12 C carbon 6 | 14 N nitrogen 7 | 16 O oxygen 8 | 19 F fluorine 9 | 20 Ne neon 10 |
| **period 3** | 23 Na sodium 11 | 24 Mg magnesium 12 | | | | | | | | | | | | | 27 Al aluminium 13 | 28 Si silicon 14 | 31 P phosphorus 15 | 32 S sulfur 16 | 35.5 Cl chlorin 17 | 40 Ar argon 18 |
| **period 4** | 39 K potassium 19 | 40 Ca calcium 20 | 45 Sc scandium 21 | 48 Ti titanium 22 | 51 V vanadium 23 | 52 Cr chromium 24 | 55 Mn manganese 25 | 56 Fe iron 26 | 59 Co cobalt 27 | 59 Ni nickel 28 | 63.5 Cu copper 29 | 65 Zn zinc 30 | | | 70 Ga gallium 31 | 73 Ge germanium 32 | 75 As arsenic 33 | 79 Se selenium 34 | 80 Br bromine 35 | 84 Kr krypton 36 |
| **period 5** | 86 Rb rubidium 37 | 88 Sr strontium 38 | 89 Y yttrium 39 | 91 Zr zirconium 40 | 93 Nb niobium 41 | 96 Mo molybdenum 42 | 98 Tc technetium 43 | 101 Ru ruthenium 44 | 103 Rh rhodium 45 | 106 Pd palladium 46 | 108 Ag silver 47 | 112 Cd cadmium 48 | | | 115 In indium 49 | 119 Sn tin 50 | 122 Sb antimony 51 | 126 Te tellurium 52 | 127 I iodine 53 | 131 Xe xenon 54 |
| **period 6** | 133 Cs caesium 55 | 137 Ba barium 56 | 139 La lanthanum 57 | 178 Hf hafnium 72 | 181 Ta tantalum 73 | 184 W tungsten 74 | 186 Re rhenium 75 | 190 Os osmium 76 | 192 Ir iridium 77 | 195 Pt platinum 78 | 197 Au gold 79 | 201 Hg mercury 80 | | | 204 Tl thallium 81 | 207 Pb lead 82 | 209 Bi bismuth 83 | 209 Po polonium 84 | 210 At astatine 85 | 222 Rn radon 86 |
| **period 7** | 223 Fr francium 87 | 226 Ra radium 88 | 227 Ac actinium 89 | 104 | 105 | 106 | 107 | 108 | 109 | 110 | 111 | 112 | | | | | | | | |

# OXFORD
## UNIVERSITY PRESS

Great Clarendon Street, Oxford OX2 6DP

Oxford University Press is a department of the University of Oxford.
It furthers the University's objective of excellence in research,
scholarship, and education by publishing worldwide in

Oxford   New York

Auckland   Cape Town   Dar es Salaam   Hong Kong   Karachi
Kuala Lumpur   Madrid   Melbourne   Mexico City   Nairobi
New Delhi   Shanghai   Taipei   Toronto

With offices in
Argentina   Austria   Brazil   Chile   Czech Republic   France   Greece
Guatemala   Hungary   Italy   Japan   Poland   Portugal   Singapore
South Korea   Switzerland   Thailand   Turkey   Ukraine   Vietnam

Oxford is a registered trade mark of Oxford University Press
in the UK and in certain other countries.

British Library Cataloguing in Publication Data.

Data available.

ISBN 978-0-19-913826-5

10 9 8 7 6 5 4 3 2 1

Printed in Great Britain by Bell and Bain Ltd, Glasgow.

Paper used in the production of this book is a natural, recyclable product
made from wood grown in sustainable forests. The manufacturing process
conforms to the environmental regulations of the country of origin.

**Acknowledgements**
Illustrations by IFA Design, Plymouth, UK, Clive Goodyer, and Q2A Media.

**Project Team acknowledgements**
These resources have been developed to support teachers and students
undertaking the OCR suite of specifications GCSE Science Twenty First Century
Science. They have been developed from the 2006 edition of the resources.
We would like to thank David Curnow and Alistair Moore and the examining
team at OCR, who produced the specifications for the Twenty First Century
Science course.

**Authors and editors of the first edition**
We thank the authors and editors of the first edition, Jenifer Burden, Peter
Campbell, Angela Hall, Andrew Hunt, Robin Millar, Caroline Shearer, and
Charles Tracy.
Many people from schools, colleges, universities, industry, and the professions
contributed to the production of the first edition of these resources. We also
acknowledge the invaluable contribution of the teachers and students in the
pilot centres.
The first edition of Twenty First Century Science was developed with support
from the Nuffield Foundation, The Salters Institute, and the Wellcome Trust.
A full list of contributors can be found in the Teacher and Technician
Resources.
The continued development of Twenty First Century Science is made possible
by generous support from:
• The Nuffield Foundation
• The Salters' Institute

MIX
Paper from
responsible sources
FSC   FSC® C007785
www.fsc.org